Michael Spivak

Brandeis University

Calculus on Manifolds

A MODERN APPROACH TO CLASSICAL THEOREMS
OF ADVANCED CALCULUS

THE BENJAMIN/CUMMINGS PUBLISHING COMPANY

Menlo Park, California • Reading, Massachusetts
London • Amsterdam • Don Mills, Ontario • Sydney

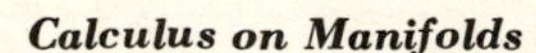

Calculus on Manifolds

A Modern Approach to Classical Theorems of Advanced Calculus

Library of Congress Catalog Card Number 66-10910
Manufactured in the United States of America
The manuscript was put into production on April 21, 1965;
this volume was published on October 26, 1965

ISBN 0-8053-9021-9
IJKLMNOP-AL-798

New York, New York 10016
67890M21098

Editors' Foreword

Mathematics has been expanding in all directions at a fabulous rate during the past half century. New fields have emerged, the diffusion into other disciplines has proceeded apace, and our knowledge of the classical areas has grown ever more profound. At the same time, one of the most striking trends in modern mathematics is the constantly increasing interrelationship between its various branches. Thus the present-day students of mathematics are faced with an immense mountain of material. In addition to the traditional areas of mathematics as presented in the traditional manner—and these presentations do abound—there are the new and often enlightening ways of looking at these traditional areas, and also the vast new areas teeming with potentialities. Much of this new material is scattered indigestibly throughout the research journals, and frequently coherently organized only in the minds or unpublished notes of the working mathematicians. And students desperately need to learn more and more of this material.

This series of brief topical booklets has been conceived as a possible means to tackle and hopefully to alleviate some of these pedagogical problems. They are being written by active research mathematicians, who can look at the latest developments, who can use these developments to clarify and condense the required material, who know what ideas to underscore and what techniques to stress. We hope that they will also serve to present to the able undergraduate an introduction to contemporary research and problems in mathematics, and that they will be sufficiently informal that the personal tastes and attitudes of the leaders in modern mathematics will shine through clearly to the readers.

The area of differential geometry is one in which recent developments have effected great changes. That part of differential geometry centered about Stokes' Theorem, sometimes called the fundamental theorem of multivariate calculus, is traditionally taught in advanced calculus courses (second or third year) and is essential in engineering and physics as well as in several current and important branches of mathematics. However, the teaching of this material has been relatively little affected by these modern developments; so the mathematicians must relearn the material in graduate school, and other scientists are frequently altogether deprived of it. Dr. Spivak's book should be a help to those who wish to see Stoke's Theorem as the modern working mathematician sees it. A student with a good course in calculus and linear algebra behind him should find this book quite accessible.

Robert Gunning
Hugo Rossi

Princeton, New Jersey
Waltham, Massachusetts
August 1965

Preface

This little book is especially concerned with those portions of "advanced calculus" in which the subtlety of the concepts and methods makes rigor difficult to attain at an elementary level. The approach taken here uses elementary versions of modern methods found in sophisticated mathematics. The formal prerequisites include only a term of linear algebra, a nodding acquaintance with the notation of set theory, and a respectable first-year calculus course (one which at least mentions the least upper bound (sup) and greatest lower bound (inf) of a set of real numbers). Beyond this a certain (perhaps latent) rapport with abstract mathematics will be found almost essential.

The first half of the book covers that simple part of advanced calculus which generalizes elementary calculus to higher dimensions. Chapter 1 contains preliminaries, and Chapters 2 and 3 treat differentiation and integration.

The remainder of the book is devoted to the study of curves, surfaces, and higher-dimensional analogues. Here the modern and classical treatments pursue quite different routes; there are, of course, many points of contact, and a significant encounter occurs in the last section. The very classical equation reproduced on the cover appears also as the last theorem of the book. This theorem (Stokes' Theorem) has had a curious history and has undergone a striking metamorphosis.

The first statement of the Theorem appears as a postscript to a letter, dated July 2, 1850, from Sir William Thomson (Lord Kelvin) to Stokes. It appeared publicly as question 8 on the Smith's Prize Examination for 1854. This competitive examination, which was taken annually by the best mathematics students at Cambridge University, was set from 1849 to 1882 by Professor Stokes; by the time of his death the result was known universally as Stokes' Theorem. At least three proofs were given by his contemporaries: Thomson published one, another appeared in Thomson and Tait's *Treatise on Natural Philosophy*, and Maxwell provided another in *Electricity and Magnetism* [13]. Since this time the name of Stokes has been applied to much more general results, which have figured so prominently in the development of certain parts of mathematics that Stokes' Theorem may be considered a case study in the value of generalization.

In this book there are three forms of Stokes' Theorem. The version known to Stokes appears in the last section, along with its inseparable companions, Green's Theorem and the Divergence Theorem. These three theorems, the classical theorems of the subtitle, are derived quite easily from a modern Stokes' Theorem which appears earlier in Chapter 5. What the classical theorems state for curves and surfaces, this theorem states for the higher-dimensional analogues (manifolds) which are studied thoroughly in the first part of Chapter 5. This study of manifolds, which could be justified solely on the basis of their importance in modern mathematics, actually involves no more effort than a careful study of curves and surfaces alone would require.

The reader probably suspects that the modern Stokes' Theorem is at least as difficult as the classical theorems derived from it. On the contrary, it is a very simple consequence of yet another version of Stokes' Theorem; this very abstract version is the final and main result of Chapter 4.

It is entirely reasonable to suppose that the difficulties so far avoided must be hidden here. Yet the proof of this theorem is, in the mathematician's sense, an utter triviality—a straightforward computation. On the other hand, even the statement of this triviality cannot be understood without a horde of difficult definitions from Chapter 4. There are good reasons why the theorems should all be easy and the definitions hard. As the evolution of Stokes' Theorem revealed, a single simple principle can masquerade as several difficult results; the proofs of many theorems involve merely stripping away the disguise. The definitions, on the other hand, serve a twofold purpose: they are rigorous replacements for vague notions, and machinery for elegant proofs. The first two sections of Chapter 4 define precisely, and prove the rules for manipulating, what are classically described as "expressions of the form" $P\,dx + Q\,dy + R\,dz$, or $P\,dx\,dy + Q\,dy\,dz + R\,dz\,dx$. Chains, defined in the third section, and partitions of unity (already introduced in Chapter 3) free our proofs from the necessity of chopping manifolds up into small pieces; they reduce questions about manifolds, where everything seems hard, to questions about Euclidean space, where everything is easy.

Concentrating the depth of a subject in the definitions is undeniably economical, but it is bound to produce some difficulties for the student. I hope the reader will be encouraged to learn Chapter 4 thoroughly by the assurance that the results will justify the effort: the classical theorems of the last section represent only a few, and by no means the most important, applications of Chapter 4; many others appear as problems, and further developments will be found by exploring the bibliography.

The problems and the bibliography both deserve a few words. Problems appear after every section and are numbered (like the theorems) within chapters. I have starred those problems whose results are used in the text, but this precaution should be unnecessary—the problems are the most important part of the book, and the reader should at least attempt them all. It was necessary to make the bibliography either very incomplete or unwieldy, since half the major branches of mathematics could legitimately be recommended as reasonable continuations of the material in the book. I have tried to make it incomplete but tempting.

Many criticisms and suggestions were offered during the writing of this book. I am particularly grateful to Richard Palais, Hugo Rossi, Robert Seeley, and Charles Stenard for their many helpful comments.

I have used this printing as an opportunity to correct many misprints and minor errors pointed out to me by indulgent readers. In addition, the material following Theorem 3-11 has been completely revised and corrected. Other important changes, which could not be incorporated in the text without excessive alteration, are listed in the Addenda at the end of the book.

Michael Spivak

Waltham, Massachusetts
March 1968

Contents

Calculus on Manifolds

1

Functions on Euclidean Space

NORM AND INNER PRODUCT

Euclidean n-space $\mathbf{R}^n$ is defined as the set of all n-tuples $(x^1, \ldots, x^n)$ of real numbers x^i (a "1-tuple of numbers" is just a number and $\mathbf{R}^1 = \mathbf{R}$, the set of all real numbers). An element of $\mathbf{R}^n$ is often called a point in $\mathbf{R}^n$, and $\mathbf{R}^1$, $\mathbf{R}^2$, $\mathbf{R}^3$ are often called the line, the plane, and space, respectively. If x denotes an element of $\mathbf{R}^n$, then x is an n-tuple of numbers, the ith one of which is denoted x^i; thus we can write

$$x = (x^1, \ldots, x^n).$$

A point in $\mathbf{R}^n$ is frequently also called a vector in $\mathbf{R}^n$, because $\mathbf{R}^n$, with $x + y = (x^1 + y^1, \ldots, x^n + y^n)$ and $ax = (ax^1, \ldots, ax^n)$, as operations, *is* a vector space (over the real numbers, of dimension n). In this vector space there is the notion of the length of a vector x, usually called the **norm** $|x|$ of x and defined by $|x| = \sqrt{(x^1)^2 + \cdots + (x^n)^2}$. If $n = 1$, then $|x|$ is the usual absolute value of x. The relation between the norm and the vector space structure of $\mathbf{R}^n$ is very important.

1-1 **Theorem.** *If $x,y \in \mathbf{R}^n$ and $a \in \mathbf{R}$, then*

(1) $|x| \geq 0$, *and* $|x| = 0$ *if and only if* $x = 0$.
(2) $|\Sigma_{i=1}^n x^i y^i| \leq |x| \cdot |y|$; *equality holds if and only if x and y are linearly dependent.*
(3) $|x + y| \leq |x| + |y|$.
(4) $|ax| = |a| \cdot |x|$.

Proof

(1) is left to the reader.
(2) If x and y are linearly dependent, equality clearly holds. If not, then $\lambda y - x \neq 0$ for all $\lambda \in \mathbf{R}$, so

$$0 < |\lambda y - x|^2 = \sum_{i=1}^n (\lambda y^i - x^i)^2$$

$$= \lambda^2 \sum_{i=1}^n (y^i)^2 - 2\lambda \sum_{i=1}^n x^i y^i + \sum_{i=1}^n (x^i)^2.$$

Therefore the right side is a quadratic equation in λ with no real solution, and its discriminant must be negative. Thus

$$4 \left(\sum_{i=1}^n x^i y^i \right)^2 - 4 \sum_{i=1}^n (x^i)^2 \cdot \sum_{i=1}^n (y^i)^2 < 0.$$

(3)
$$\begin{aligned} |x + y|^2 &= \Sigma_{i=1}^n (x^i + y^i)^2 \\ &= \Sigma_{i=1}^n (x^i)^2 + \Sigma_{i=1}^n (y^i)^2 + 2\Sigma_{i=1}^n x^i y^i \\ &\leq |x|^2 + |y|^2 + 2|x| \cdot |y| \qquad \text{by (2)} \\ &= (|x| + |y|)^2. \end{aligned}$$

(4) $|ax| = \sqrt{\Sigma_{i=1}^n (ax^i)^2} = \sqrt{a^2 \Sigma_{i=1}^n (x^i)^2} = |a| \cdot |x|$. ▮

The quantity $\Sigma_{i=1}^n x^i y^i$ which appears in (2) is called the **inner product** of x and y and denoted $\langle x,y \rangle$. The most important properties of the inner product are the following.

1-2 **Theorem.** *If x, x_1, x_2 and y, y_1, y_2 are vectors in $\mathbf{R}^n$ and $a \in \mathbf{R}$, then*

(1) $\langle x,y \rangle = \langle y,x \rangle$ (*symmetry*).

(2) $\langle ax,y\rangle = \langle x,ay\rangle = a\langle x,y\rangle$ *(bilinearity)*.
$\langle x_1 + x_2, y\rangle = \langle x_1,y\rangle + \langle x_2,y\rangle$
$\langle x, y_1 + y_2\rangle = \langle x,y_1\rangle + \langle x,y_2\rangle$

(3) $\langle x,x\rangle \geq 0$, *and* $\langle x,x\rangle = 0$ *if and only if* $x = 0$ *(positive definiteness)*.

(4) $|x| = \sqrt{\langle x,x\rangle}$.

(5) $\langle x,y\rangle = \dfrac{|x+y|^2 - |x-y|^2}{4}$ *(polarization identity)*.

Proof

(1) $\langle x,y\rangle = \Sigma_{i=1}^n x^i y^i = \Sigma_{i=1}^n y^i x^i = \langle y,x\rangle$.

(2) By (1) it suffices to prove

$$\langle ax,y\rangle = a\langle x,y\rangle,$$
$$\langle x_1 + x_2, y\rangle = \langle x_1,y\rangle + \langle x_2,y\rangle.$$

These follow from the equations

$$\langle ax,y\rangle = \sum_{i=1}^n (ax^i)y^i = a\sum_{i=1}^n x^i y^i = a\langle x,y\rangle,$$
$$\langle x_1 + x_2, y\rangle = \sum_{i=1}^n (x_1{}^i + x_2{}^i)y^i = \sum_{i=1}^n x_1{}^i y^i + \sum_{i=1}^n x_2{}^i y^i$$
$$= \langle x_1,y\rangle + \langle x_2,y\rangle.$$

(3) and (4) are left to the reader.

(5) $\dfrac{|x+y|^2 - |x-y|^2}{4}$

$= \frac{1}{4}[\langle x+y, x+y\rangle - \langle x-y, x-y\rangle]$ by (4)
$= \frac{1}{4}[\langle x,x\rangle + 2\langle x,y\rangle + \langle y,y\rangle - (\langle x,x\rangle - 2\langle x,y\rangle + \langle y,y\rangle)]$
$= \langle x,y\rangle$. ▮

We conclude this section with some important remarks about notation. The vector $(0, \ldots, 0)$ will usually be denoted simply 0. The **usual basis** of $\mathbf{R}^n$ is $e_1, \ldots, e_n$, where $e_i = (0, \ldots, 1, \ldots, 0)$, with the 1 in the ith place. If $T: \mathbf{R}^n \to \mathbf{R}^m$ is a linear transformation, the matrix of T with respect to the usual bases of $\mathbf{R}^n$ and $\mathbf{R}^m$ is the $m \times n$ matrix $A = (a_{ij})$, where $T(e_i) = \Sigma_{j=1}^m a_{ji} e_j$ —the coefficients of $T(e_i)$

appear in the ith *column* of the matrix. If $S\colon \mathbf{R}^m \to \mathbf{R}^p$ has the $p \times m$ matrix B, then $S \circ T$ has the $p \times n$ matrix BA [here $S \circ T(x) = S(T(x))$; most books on linear algebra denote $S \circ T$ simply ST]. To find $T(x)$ one computes the $m \times 1$ matrix

$$\begin{pmatrix} y^1 \\ \cdot \\ \cdot \\ \cdot \\ y^m \end{pmatrix} = \begin{pmatrix} a_{11}, & \dots & ,a_{1n} \\ \cdot & & \cdot \\ \cdot & & \cdot \\ \cdot & & \cdot \\ a_{m1}, & \dots & ,a_{mn} \end{pmatrix} \cdot \begin{pmatrix} x^1 \\ \cdot \\ \cdot \\ \cdot \\ x^n \end{pmatrix};$$

then $T(x) = (y^1, \dots ,y^m)$. One notational convention greatly simplifies many formulas: if $x \in \mathbf{R}^n$ and $y \in \mathbf{R}^m$, then (x,y) denotes

$$(x^1, \dots ,x^n,y^1, \dots ,y^m) \in \mathbf{R}^{n+m}.$$

Problems. **1-1.*** Prove that $|x| \le \sum_{i=1}^n |x^i|$.

1-2. When does equality hold in Theorem 1-1(3)? *Hint:* Re-examine the proof; the answer is not "when x and y are linearly dependent."

1-3. Prove that $|x - y| \le |x| + |y|$. When does equality hold?

1-4. Prove that $\big|\,|x| - |y|\,\big| \le |x - y|$.

1-5. The quantity $|y - x|$ is called the **distance** between x and y. Prove and interpret geometrically the "triangle inequality": $|z - x| \le |z - y| + |y - x|$.

1-6. Let f and g be integrable on $[a,b]$.

(a) Prove that $\left|\int_a^b f \cdot g\right| \le (\int_a^b f^2)^{\frac{1}{2}} \cdot (\int_a^b g^2)^{\frac{1}{2}}$. *Hint:* Consider separately the cases $0 = \int_a^b (f - \lambda g)^2$ for some $\lambda \in \mathbf{R}$ and $0 < \int_a^b (f - \lambda g)^2$ for all $\lambda \in \mathbf{R}$.

(b) If equality holds, must $f = \lambda g$ for some $\lambda \in \mathbf{R}$? What if f and g are continuous?

(c) Show that Theorem 1-1(2) is a special case of (a).

1-7. A linear transformation $T\colon \mathbf{R}^n \to \mathbf{R}^n$ is **norm preserving** if $|T(x)| = |x|$, and **inner product preserving** if $\langle Tx,Ty \rangle = \langle x,y \rangle$.

(a) Prove that T is norm preserving if and only if T is inner-product preserving.

(b) Prove that such a linear transformation T is 1-1 and T^{-1} is of the same sort.

1-8. If $x,y \in \mathbf{R}^n$ are non-zero, the **angle** between x and y, denoted $\angle(x,y)$, is defined as $\arccos(\langle x,y \rangle/|x| \cdot |y|)$, which makes sense by Theorem 1-1(2). The linear transformation T is **angle preserving** if T is 1-1, and for $x,y \ne 0$ we have $\angle(Tx,Ty) = \angle(x,y)$.

(a) Prove that if T is norm preserving, then T is angle preserving.

(b) If there is a basis $x_1, \ldots, x_n$ of $\mathbf{R}^n$ and numbers $\lambda_1, \ldots, \lambda_n$ such that $Tx_i = \lambda_i x_i$, prove that T is angle preserving if and only if all $|\lambda_i|$ are equal.

(c) What are all angle preserving $T: \mathbf{R}^n \to \mathbf{R}^n$?

1-9. If $0 \leq \theta < \pi$, let $T: \mathbf{R}^2 \to \mathbf{R}^2$ have the matrix $\begin{pmatrix} \cos\theta, & \sin\theta \\ -\sin\theta, & \cos\theta \end{pmatrix}$. Show that T is angle preserving and if $x \neq 0$, then $\angle(x,Tx) = \theta$.

1-10.* If $T: \mathbf{R}^m \to \mathbf{R}^n$ is a linear transformation, show that there is a number M such that $|T(h)| \leq M|h|$ for $h \in \mathbf{R}^m$. *Hint:* Estimate $|T(h)|$ in terms of $|h|$ and the entries in the matrix of T.

1-11. If $x,y \in \mathbf{R}^n$ and $z,w \in \mathbf{R}^m$, show that $\langle (x,z),(y,w) \rangle = \langle x,y \rangle + \langle z,w \rangle$ and $|(x,z)| = \sqrt{|x|^2 + |z|^2}$. Note that (x,z) and (y,w) denote points in $\mathbf{R}^{n+m}$.

1-12.* Let $(\mathbf{R}^n)^*$ denote the dual space of the vector space $\mathbf{R}^n$. If $x \in \mathbf{R}^n$, define $\varphi_x \in (\mathbf{R}^n)^*$ by $\varphi_x(y) = \langle x,y \rangle$. Define $T: \mathbf{R}^n \to (\mathbf{R}^n)^*$ by $T(x) = \varphi_x$. Show that T is a 1-1 linear transformation and conclude that every $\varphi \in (\mathbf{R}^n)^*$ is φ_x for a unique $x \in \mathbf{R}^n$.

1-13.* If $x,y \in \mathbf{R}^n$, then x and y are called **perpendicular** (or **orthogonal**) if $\langle x,y \rangle = 0$. If x and y are perpendicular, prove that $|x + y|^2 = |x|^2 + |y|^2$.

SUBSETS OF EUCLIDEAN SPACE

The closed interval $[a,b]$ has a natural analogue in $\mathbf{R}^2$. This is the **closed rectangle** $[a,b] \times [c,d]$, defined as the collection of all pairs (x,y) with $x \in [a,b]$ and $y \in [c,d]$. More generally, if $A \subset \mathbf{R}^m$ and $B \subset \mathbf{R}^n$, then $A \times B \subset \mathbf{R}^{m+n}$ is defined as the set of all $(x,y) \in \mathbf{R}^{m+n}$ with $x \in A$ and $y \in B$. In particular, $\mathbf{R}^{m+n} = \mathbf{R}^m \times \mathbf{R}^n$. If $A \subset \mathbf{R}^m$, $B \subset \mathbf{R}^n$, and $C \subset \mathbf{R}^p$, then $(A \times B) \times C = A \times (B \times C)$, and both of these are denoted simply $A \times B \times C$; this convention is extended to the product of any number of sets. The set $[a_1,b_1] \times \cdots \times [a_n,b_n] \subset \mathbf{R}^n$ is called a **closed rectangle** in $\mathbf{R}^n$, while the set $(a_1,b_1) \times \cdots \times (a_n,b_n) \subset \mathbf{R}^n$ is called an **open rectangle.** More generally a set $U \subset \mathbf{R}^n$ is called **open** (Figure 1-1) if for each $x \in U$ there is an open rectangle A such that $x \in A \subset U$.

A subset C of $\mathbf{R}^n$ is **closed** if $\mathbf{R}^n - C$ is open. For example, if C contains only finitely many points, then C is closed.

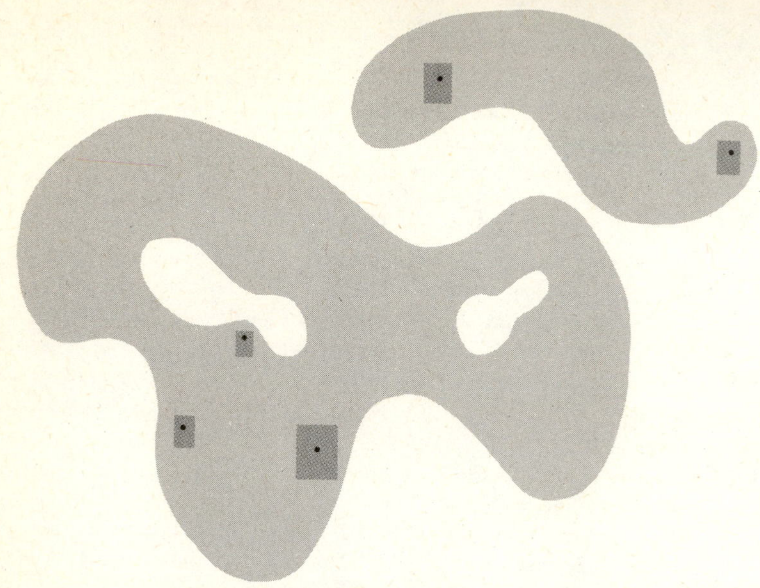

FIGURE 1-1

The reader should supply the proof that a closed rectangle in $\mathbf{R}^n$ is indeed a closed set.

If $A \subset \mathbf{R}^n$ and $x \in \mathbf{R}^n$, then one of three possibilities must hold (Figure 1-2):

1. There is an open rectangle B such that $x \in B \subset A$.
2. There is an open rectangle B such that $x \in B \subset \mathbf{R}^n - A$.
3. If B is any open rectangle with $x \in B$, then B contains points of both A and $\mathbf{R}^n - A$.

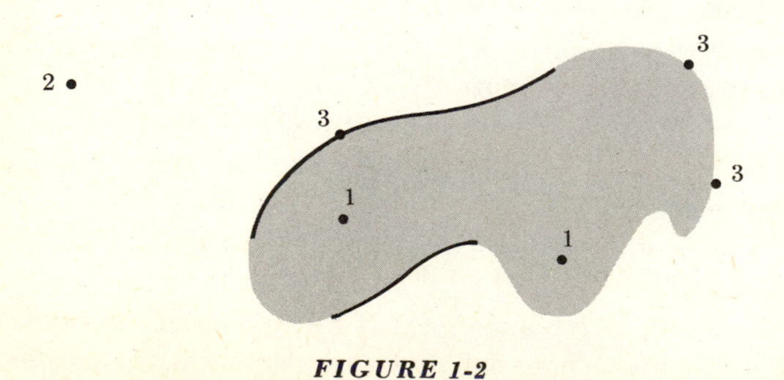

FIGURE 1-2

Those points satisfying (1) constitute the **interior** of A, those satisfying (2) the **exterior** of A, and those satisfying (3) the **boundary** of A. Problems 1-16 to 1-18 show that these terms may sometimes have unexpected meanings.

It is not hard to see that the interior of any set A is open, and the same is true for the exterior of A, which is, in fact, the interior of $\mathbf{R}^n - A$. Thus (Problem 1-14) their union is open, and what remains, the boundary, must be closed.

A collection $\mathcal{O}$ of open sets is an **open cover** of A (or, briefly, **covers** A) if every point $x \in A$ is in some open set in the collection $\mathcal{O}$. For example, if $\mathcal{O}$ is the collection of all open intervals $(a, a + 1)$ for $a \in \mathbf{R}$, then $\mathcal{O}$ is a cover of $\mathbf{R}$. Clearly no finite number of the open sets in $\mathcal{O}$ will cover $\mathbf{R}$ or, for that matter, any unbounded subset of $\mathbf{R}$. A similar situation can also occur for bounded sets. If $\mathcal{O}$ is the collection of all open intervals $(1/n, 1 - 1/n)$ for all integers $n > 1$, then $\mathcal{O}$ is an open cover of $(0,1)$, but again no finite collection of sets in $\mathcal{O}$ will cover $(0,1)$. Although this phenomenon may not appear particularly scandalous, sets for which this state of affairs cannot occur are of such importance that they have received a special designation: a set A is called **compact** if every open cover $\mathcal{O}$ contains a finite subcollection of open sets which also covers A.

A set with only finitely many points is obviously compact and so is the infinite set A which contains 0 and the numbers $1/n$ for all integers n (reason: if $\mathcal{O}$ is a cover, then $0 \in U$ for some open set U in $\mathcal{O}$; there are only finitely many other points of A not in U, each requiring at most one more open set).

Recognizing compact sets is greatly simplified by the following results, of which only the first has any depth (i.e., uses any facts about the real numbers).

1-3 Theorem (Heine-Borel). *The closed interval* $[a,b]$ *is compact.*

Proof. If $\mathcal{O}$ is an open cover of $[a,b]$, let

$$A = \{x \colon a \le x \le b \text{ and } [a,x] \text{ is covered by some finite number of open sets in } \mathcal{O}\}.$$

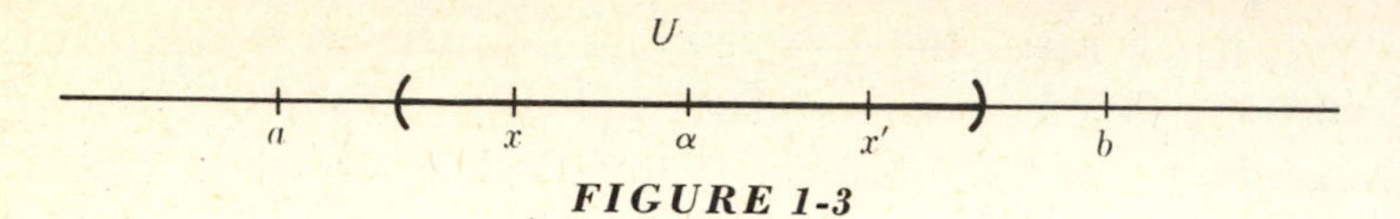

FIGURE 1-3

Note that $a \in A$ and that A is clearly bounded above (by b). We would like to show that $b \in A$. This is done by proving two things about α = least upper bound of A; namely, (1) $\alpha \in A$ and (2) $b = \alpha$.

Since $\mathcal{O}$ is a cover, $\alpha \in U$ for some U in $\mathcal{O}$. Then all points in some interval to the left of α are also in U (see Figure 1-3). Since α is the least upper bound of A, there is an x in this interval such that $x \in A$. Thus $[a,x]$ is covered by some finite number of open sets of $\mathcal{O}$, while $[x,\alpha]$ is covered by the single set U. Hence $[a,\alpha]$ is covered by a finite number of open sets of $\mathcal{O}$, and $\alpha \in A$. This proves (1).

To prove that (2) is true, suppose instead that $\alpha < b$. Then there is a point x' between α and b such that $[\alpha,x'] \subset U$. Since $\alpha \in A$, the interval $[a,\alpha]$ is covered by finitely many open sets of $\mathcal{O}$, while $[\alpha,x']$ is covered by U. Hence $x' \in A$, contradicting the fact that α is an upper bound of A. ▌

If $B \subset \mathbf{R}^m$ is compact and $x \in \mathbf{R}^n$, it is easy to see that $\{x\} \times B \subset \mathbf{R}^{n+m}$ is compact. However, a much stronger assertion can be made.

1-4 Theorem. *If B is compact and $\mathcal{O}$ is an open cover of $\{x\} \times B$, then there is an open set $U \subset \mathbf{R}^n$ containing x such that $U \times B$ is covered by a finite number of sets in $\mathcal{O}$.*

Proof. Since $\{x\} \times B$ is compact, we can assume at the outset that $\mathcal{O}$ is finite, and we need only find the open set U such that $U \times B$ is covered by $\mathcal{O}$.

For each $y \in B$ the point (x,y) is in some open set W in $\mathcal{O}$. Since W is open, we have $(x,y) \in U_y \times V_y \subset W$ for some open rectangle $U_y \times V_y$. The sets V_y cover the compact set B, so a finite number $V_{y_1}, \ldots, V_{y_k}$ also cover B. Let $U = U_{y_1} \cap \cdots \cap U_{y_k}$. Then if $(x',y') \in U \times B$, we have

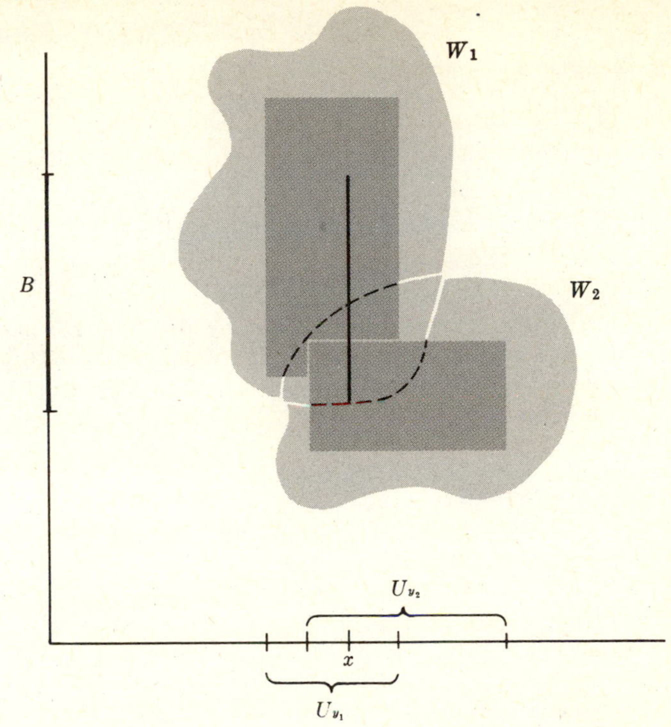

FIGURE 1-4

$y' \in V_{y_i}$ for some i (Figure 1-4), and certainly $x' \in U_{y_i}$. Hence $(x',y') \in U_{y_i} \times V_{y_i}$, which is contained in some W in $\mathcal{O}$. ∎

1-5 Corollary. *If $A \subset \mathbf{R}^n$ and $B \subset \mathbf{R}^m$ are compact, then $A \times B \subset \mathbf{R}^{n+m}$ is compact.*

Proof. If $\mathcal{O}$ is an open cover of $A \times B$, then $\mathcal{O}$ covers $\{x\} \times B$ for each $x \in A$. By Theorem 1-4 there is an open set U_x containing x such that $U_x \times B$ is covered by finitely many sets in $\mathcal{O}$. Since A is compact, a finite number $U_{x_1}, \ldots, U_{x_n}$ of the U_x cover A. Since finitely many sets in $\mathcal{O}$ cover each $U_{x_i} \times B$, finitely many cover all of $A \times B$. ∎

1-6 Corollary. *$A_1 \times \cdots \times A_k$ is compact if each A_i is. In particular, a closed rectangle in $\mathbf{R}^k$ is compact.*

1-7 Corollary. *A closed bounded subset of* $\mathbf{R}^n$ *is compact.* (The converse is also true (Problem 1-20).)

Proof. If $A \subset \mathbf{R}^n$ is closed and bounded, then $A \subset B$ for some closed rectangle B. If $\mathcal{O}$ is an open cover of A, then $\mathcal{O}$ together with $\mathbf{R}^n - A$ is an open cover of B. Hence a finite number $U_1, \ldots, U_n$ of sets in $\mathcal{O}$, together with $\mathbf{R}^n - A$ perhaps, cover B. Then $U_1, \ldots, U_n$ cover A. ▌

Problems. **1-14.*** Prove that the union of any (even infinite) number of open sets is open. Prove that the intersection of two (and hence of finitely many) open sets is open. Give a counterexample for infinitely many open sets.

1-15. Prove that $\{x \in \mathbf{R}^n: |x - a| < r\}$ is open (see also Problem 1-27).

1-16. Find the interior, exterior, and boundary of the sets

$$\{x \in \mathbf{R}^n: |x| \leq 1\}$$
$$\{x \in \mathbf{R}^n: |x| = 1\}$$
$$\{x \in \mathbf{R}^n: \text{each } x^i \text{ is rational}\}.$$

1-17. Construct a set $A \subset [0,1] \times [0,1]$ such that A contains at most one point on each horizontal and each vertical line but boundary $A = [0,1] \times [0,1]$. *Hint:* It suffices to ensure that A contains points in each quarter of the square $[0,1] \times [0,1]$ and also in each sixteenth, etc.

1-18. If $A \subset [0,1]$ is the union of open intervals (a_i,b_i) such that each rational number in $(0,1)$ is contained in some (a_i,b_i), show that boundary $A = [0,1] - A$.

1-19.* If A is a closed set that contains every rational number $r \in [0,1]$, show that $[0,1] \subset A$.

1-20. Prove the converse of Corollary 1-7: A compact subset of $\mathbf{R}^n$ is closed and bounded (see also Problem 1-28).

1-21.* (a) If A is closed and $x \notin A$, prove that there is a number $d > 0$ such that $|y - x| \geq d$ for all $y \in A$.

(b) If A is closed, B is compact, and $A \cap B = \varnothing$, prove that there is $d > 0$ such that $|y - x| \geq d$ for all $y \in A$ and $x \in B$. *Hint:* For each $b \in B$ find an open set U containing b such that this relation holds for $x \in U \cap B$.

(c) Give a counterexample in $\mathbf{R}^2$ if A and B are closed but neither is compact.

1-22.* If U is open and $C \subset U$ is compact, show that there is a compact set D such that $C \subset$ interior D and $D \subset U$.

FUNCTIONS AND CONTINUITY

A **function** from $\mathbf{R}^n$ to $\mathbf{R}^m$ (sometimes called a (vector-valued) function of n variables) is a rule which associates to each point in $\mathbf{R}^n$ some point in $\mathbf{R}^m$; the point a function f associates to x is denoted $f(x)$. We write $f\colon \mathbf{R}^n \to \mathbf{R}^m$ (read "f takes $\mathbf{R}^n$ into $\mathbf{R}^m$" or "f, taking $\mathbf{R}^n$ into $\mathbf{R}^m$," depending on context) to indicate that $f(x) \in \mathbf{R}^m$ is defined for $x \in \mathbf{R}^n$. The notation $f\colon A \to \mathbf{R}^m$ indicates that $f(x)$ is defined only for x in the set A, which is called the **domain** of f. If $B \subset A$, we define $f(B)$ as the set of all $f(x)$ for $x \in B$, and if $C \subset \mathbf{R}^m$ we define $f^{-1}(C) = \{x \in A : f(x) \in C\}$. The notation $f\colon A \to B$ indicates that $f(A) \subset B$.

A convenient representation of a function $f\colon \mathbf{R}^2 \to \mathbf{R}$ may be obtained by drawing a picture of its graph, the set of all 3-tuples of the form $(x,y,f(x,y))$, which is actually a figure in 3-space (see, e.g., Figures 2-1 and 2-2 of Chapter 2).

If $f,g\colon \mathbf{R}^n \to \mathbf{R}$, the functions $f + g$, $f - g$, $f \cdot g$, and f/g are defined precisely as in the one-variable case. If $f\colon A \to \mathbf{R}^m$ and $g\colon B \to \mathbf{R}^p$, where $B \subset \mathbf{R}^m$, then the **composition** $g \circ f$ is defined by $g \circ f(x) = g(f(x))$; the domain of $g \circ f$ is $A \cap f^{-1}(B)$. If $f\colon A \to \mathbf{R}^m$ is 1-1, that is, if $f(x) \neq f(y)$ when $x \neq y$, we define $f^{-1}\colon f(A) \to \mathbf{R}^n$ by the requirement that $f^{-1}(z)$ is the unique $x \in A$ with $f(x) = z$.

A function $f\colon A \to \mathbf{R}^m$ determines m **component functions** $f^1, \ldots ,f^m\colon A \to \mathbf{R}$ by $f(x) = (f^1(x), \ldots ,f^m(x))$. If conversely, m functions $g_1, \ldots ,g_m\colon A \to \mathbf{R}$ are given, there is a unique function $f\colon A \to \mathbf{R}^m$ such that $f^i = g_i$, namely $f(x) = (g_1(x), \ldots ,g_m(x))$. This function f will be denoted $(g_1, \ldots ,g_m)$, so that we always have $f = (f^1, \ldots ,f^m)$. If $\pi\colon \mathbf{R}^n \to \mathbf{R}^n$ is the identity function, $\pi(x) = x$, then $\pi^i(x) = x^i$; the function π^i is called the ith **projection function.**

The notation $\lim_{x\to a} f(x) = b$ means, as in the one-variable case, that we can get $f(x)$ as close to b as desired, by choosing x sufficiently close to, but not equal to, a. In mathematical terms this means that for every number $\varepsilon > 0$ there is a number $\delta > 0$ such that $|f(x) - b| < \varepsilon$ for all x in the domain of f which satisfy $0 < |x - a| < \delta$. A function $f: A \to \mathbf{R}^m$ is called **continuous** at $a \in A$ if $\lim_{x \to a} f(x) = f(a)$, and f is simply called continuous if it is continuous at each $a \in A$. One of the pleasant surprises about the concept of continuity is that it can be defined without using limits. It follows from the next theorem that $f: \mathbf{R}^n \to \mathbf{R}^m$ is continuous if and only if $f^{-1}(U)$ is open whenever $U \subset \mathbf{R}^m$ is open; if the domain of f is not all of $\mathbf{R}^n$, a slightly more complicated condition is needed.

1-8 Theorem. *If $A \subset \mathbf{R}^n$, a function $f: A \to \mathbf{R}^m$ is continuous if and only if for every open set $U \subset \mathbf{R}^m$ there is some open set $V \subset \mathbf{R}^n$ such that $f^{-1}(U) = V \cap A$.*

Proof. Suppose f is continuous. If $a \in f^{-1}(U)$, then $f(a) \in U$. Since U is open, there is an open rectangle B with $f(a) \in B \subset U$. Since f is continuous at a, we can ensure that $f(x) \in B$, provided we choose x in some sufficiently small rectangle C containing a. Do this for each $a \in f^{-1}(U)$ and let V be the union of all such C. Clearly $f^{-1}(U) = V \cap A$. The converse is similar and is left to the reader. ▌

The following consequence of Theorem 1-8 is of great importance.

1-9 Theorem. *If $f: A \to \mathbf{R}^m$ is continuous, where $A \subset \mathbf{R}^n$, and A is compact, then $f(A) \subset \mathbf{R}^m$ is compact.*

Proof. Let $\mathcal{O}$ be an open cover of $f(A)$. For each open set U in $\mathcal{O}$ there is an open set V_U such that $f^{-1}(U) = V_U \cap A$. The collection of all V_U is an open cover of A. Since A is compact, a finite number $V_{U_1}, \ldots, V_{U_n}$ cover A. Then $U_1, \ldots, U_n$ cover $f(A)$. ▌

If $f: A \to \mathbf{R}$ is bounded, the extent to which f fails to be continuous at $a \in A$ can be measured in a precise way. For $\delta > 0$ let

$$M(a,f,\delta) = \sup\{f(x) : x \in A \text{ and } |x - a| < \delta\},$$
$$m(a,f,\delta) = \inf\{f(x) : x \in A \text{ and } |x - a| < \delta\}.$$

The **oscillation** $o(f,a)$ of f at a is defined by $o(f,a) = \lim_{\delta \to 0}[M(a,f,\delta) - m(a,f,\delta)]$. This limit always exists, since $M(a,f,\delta) - m(a,f,\delta)$ decreases as δ decreases. There are two important facts about $o(f,a)$.

1-10 Theorem. *The bounded function f is continuous at a if and only if $o(f,a) = 0$.*

Proof. Let f be continuous at a. For every number $\varepsilon > 0$ we can choose a number $\delta > 0$ so that $|f(x) - f(a)| < \varepsilon$ for all $x \in A$ with $|x - a| < \delta$; thus $M(a,f,\delta) - m(a,f,\delta) \le 2\varepsilon$. Since this is true for every ε, we have $o(f,a) = 0$. The converse is similar and is left to the reader. ▮

1-11 Theorem. *Let $A \subset \mathbf{R}^n$ be closed. If $f\colon A \to \mathbf{R}$ is any bounded function, and $\varepsilon > 0$, then $\{x \in A\colon o(f,x) \ge \varepsilon\}$ is closed.*

Proof. Let $B = \{x \in A\colon o(f,x) \ge \varepsilon\}$. We wish to show that $\mathbf{R}^n - B$ is open. If $x \in \mathbf{R}^n - B$, then either $x \notin A$ or else $x \in A$ and $o(f,x) < \varepsilon$. In the first case, since A is closed, there is an open rectangle C containing x such that $C \subset \mathbf{R}^n - A \subset \mathbf{R}^n - B$. In the second case there is a $\delta > 0$ such that $M(x,f,\delta) - m(x,f,\delta) < \varepsilon$. Let C be an open rectangle containing x such that $|x - y| < \delta$ for all $y \in C$. Then if $y \in C$ there is a δ_1 such that $|x - z| < \delta$ for all z satisfying $|z - y| < \delta_1$. Thus $M(y,f,\delta_1) - m(y,f,\delta_1) < \varepsilon$, and consequently $o(y,f) < \varepsilon$. Therefore $C \subset \mathbf{R}^n - B$. ▮

Problems. **1-23.** If $f\colon A \to \mathbf{R}^m$ and $a \in A$, show that $\lim_{x \to a} f(x) = b$ if and only if $\lim_{x \to a} f^i(x) = b^i$ for $i = 1, \ldots, m$.

1-24. Prove that $f\colon A \to \mathbf{R}^m$ is continuous at a if and only if each f^i is.

1-25. Prove that a linear transformation $T\colon \mathbf{R}^n \to \mathbf{R}^m$ is continuous. *Hint:* Use Problem 1-10.

1-26. Let $A = \{(x,y) \in \mathbf{R}^2\colon x > 0 \text{ and } 0 < y < x^2\}$.

(a) Show that every straight line through $(0,0)$ contains an interval around $(0,0)$ which is in $\mathbf{R}^2 - A$.

(b) Define $f\colon \mathbf{R}^2 \to \mathbf{R}$ by $f(x) = 0$ if $x \notin A$ and $f(x) = 1$ if $x \in A$. For $h \in \mathbf{R}^2$ define $g_h\colon \mathbf{R} \to \mathbf{R}$ by $g_h(t) = f(th)$. Show that each g_h is continuous at 0, but f is not continuous at $(0,0)$.

1-27. Prove that $\{x \in \mathbf{R}^n: |x - a| < r\}$ is open by considering the function $f: \mathbf{R}^n \to \mathbf{R}$ with $f(x) = |x - a|$.

1-28. If $A \subset \mathbf{R}^n$ is not closed, show that there is a continuous function $f: A \to \mathbf{R}$ which is unbounded. *Hint:* If $x \in \mathbf{R}^n - A$ but $x \notin$ interior $(\mathbf{R}^n - A)$, let $f(y) = 1/|y - x|$.

1-29. If A is compact, prove that every continuous function $f: A \to \mathbf{R}$ takes on a maximum and a minimum value.

1-30. Let $f: [a,b] \to \mathbf{R}$ be an increasing function. If $x_1, \ldots, x_n \in [a,b]$ are distinct, show that $\sum_{i=1}^{n} o(f,x_i) < f(b) - f(a)$.

2

Differentiation

BASIC DEFINITIONS

Recall that a function $f\colon \mathbf{R} \to \mathbf{R}$ is differentiable at $a \in \mathbf{R}$ if there is a number $f'(a)$ such that

$$(1)\ \lim_{h \to 0} \frac{f(a+h) - f(a)}{h} = f'(a).$$

This equation certainly makes no sense in the general case of a function $f\colon \mathbf{R}^n \to \mathbf{R}^m$, but can be reformulated in a way that does. If $\lambda\colon \mathbf{R} \to \mathbf{R}$ is the linear transformation defined by $\lambda(h) = f'(a) \cdot h$, then equation (1) is equivalent to

$$(2)\ \lim_{h \to 0} \frac{f(a+h) - f(a) - \lambda(h)}{h} = 0.$$

Equation (2) is often interpreted as saying that $\lambda + f(a)$ is a good approximation to f at a (see Problem 2-9). Henceforth we focus our attention on the linear transformation λ and reformulate the definition of differentiability as follows.

A function $f\colon \mathbf{R} \to \mathbf{R}$ is differentiable at $a \in \mathbf{R}$ if there is a linear transformation $\lambda\colon \mathbf{R} \to \mathbf{R}$ such that

$$\lim_{h \to 0} \frac{f(a+h) - f(a) - \lambda(h)}{h} = 0.$$

In this form the definition has a simple generalization to higher dimensions:

A function $f\colon \mathbf{R}^n \to \mathbf{R}^m$ is **differentiable** at $a \in \mathbf{R}^n$ if there is a linear transformation $\lambda\colon \mathbf{R}^n \to \mathbf{R}^m$ such that

$$\lim_{h \to 0} \frac{|f(a+h) - f(a) - \lambda(h)|}{|h|} = 0.$$

Note that h is a point of $\mathbf{R}^n$ and $f(a+h) - f(a) - \lambda(h)$ a point of $\mathbf{R}^m$, so the norm signs are essential. The linear transformation λ is denoted $Df(a)$ and called the **derivative** of f at a. The justification for the phrase "*the* linear transformation λ" is

2-1 Theorem. *If $f\colon \mathbf{R}^n \to \mathbf{R}^m$ is differentiable at $a \in \mathbf{R}^n$, there is a* unique *linear transformation $\lambda\colon \mathbf{R}^n \to \mathbf{R}^m$ such that*

$$\lim_{h \to 0} \frac{|f(a+h) - f(a) - \lambda(h)|}{|h|} = 0.$$

Proof. Suppose $\mu\colon \mathbf{R}^n \to \mathbf{R}^m$ satisfies

$$\lim_{h \to 0} \frac{|f(a+h) - f(a) - \mu(h)|}{|h|} = 0.$$

If $d(h) = f(a+h) - f(a)$, then

$$\begin{aligned}
\lim_{h \to 0} \frac{|\lambda(h) - \mu(h)|}{|h|} &= \lim_{h \to 0} \frac{|\lambda(h) - d(h) + d(h) - \mu(h)|}{|h|} \\
&\leq \lim_{h \to 0} \frac{|\lambda(h) - d(h)|}{|h|} + \lim_{h \to 0} \frac{|d(h) - \mu(h)|}{|h|} \\
&= 0.
\end{aligned}$$

If $x \in \mathbf{R}^n$, then $tx \to 0$ as $t \to 0$. Hence for $x \neq 0$ we have

$$0 = \lim_{t \to 0} \frac{|\lambda(tx) - \mu(tx)|}{|tx|} = \frac{|\lambda(x) - \mu(x)|}{|x|}.$$

Therefore $\lambda(x) = \mu(x)$. ▌

We shall later discover a simple way of finding $Df(a)$. For the moment let us consider the function $f: \mathbf{R}^2 \to \mathbf{R}$ defined by $f(x,y) = \sin x$. Then $Df(a,b) = \lambda$ satisfies $\lambda(x,y) = (\cos a) \cdot x$. To prove this, note that

$$\lim_{(h,k)\to 0} \frac{|f(a+h, b+k) - f(a,b) - \lambda(h,k)|}{|(h,k)|} = \lim_{(h,k)\to 0} \frac{|\sin(a+h) - \sin a - (\cos a) \cdot h|}{|(h,k)|}.$$

Since $\sin'(a) = \cos a$, we have

$$\lim_{h\to 0} \frac{|\sin(a+h) - \sin a - (\cos a) \cdot h|}{|h|} = 0.$$

Since $|(h,k)| \geq |h|$, it is also true that

$$\lim_{h\to 0} \frac{|\sin(a+h) - \sin a - (\cos a) \cdot h|}{|(h,k)|} = 0.$$

It is often convenient to consider the matrix of $Df(a)$: $\mathbf{R}^n \to \mathbf{R}^m$ with respect to the usual bases of $\mathbf{R}^n$ and $\mathbf{R}^m$. This $m \times n$ matrix is called the **Jacobian matrix** of f at a, and denoted $f'(a)$. If $f(x,y) = \sin x$, then $f'(a,b) = (\cos a, 0)$. If $f: \mathbf{R} \to \mathbf{R}$, then $f'(a)$ is a 1×1 matrix whose single entry is the number which is denoted $f'(a)$ in elementary calculus.

The definition of $Df(a)$ could be made if f were defined only in some open set containing a. Considering only functions defined on $\mathbf{R}^n$ streamlines the statement of theorems and produces no real loss of generality. It is convenient to define a function $f: \mathbf{R}^n \to \mathbf{R}^m$ to be differentiable on A if f is differentiable at a for each $a \in A$. If $f: A \to \mathbf{R}^m$, then f is called differentiable if f can be extended to a differentiable function on some open set containing A.

Problems. **2-1.*** Prove that if $f: \mathbf{R}^n \to \mathbf{R}^m$ is differentiable at $a \in \mathbf{R}^n$, then it is continuous at a. *Hint:* Use Problem 1-10.

2-2. A function $f: \mathbf{R}^2 \to \mathbf{R}$ is **independent of the second variable** if for each $x \in \mathbf{R}$ we have $f(x,y_1) = f(x,y_2)$ for all $y_1,y_2 \in \mathbf{R}$. Show that f is independent of the second variable if and only if there is a function $g: \mathbf{R} \to \mathbf{R}$ such that $f(x,y) = g(x)$. What is $f'(a,b)$ in terms of g'?

2-3. Define when a function $f: \mathbf{R}^2 \to \mathbf{R}$ is independent of the first variable and find $f'(a,b)$ for such f. Which functions are independent of the first variable and also of the second variable?

2-4. Let g be a continuous real-valued function on the unit circle $\{x \in \mathbf{R}^2: |x| = 1\}$ such that $g(0,1) = g(1,0) = 0$ and $g(-x) = -g(x)$. Define $f: \mathbf{R}^2 \to \mathbf{R}$ by

$$f(x) = \begin{cases} |x| \cdot g\left(\dfrac{x}{|x|}\right) & x \neq 0, \\ 0 & x = 0. \end{cases}$$

(a) If $x \in \mathbf{R}^2$ and $h: \mathbf{R} \to \mathbf{R}$ is defined by $h(t) = f(tx)$, show that h is differentiable.

(b) Show that f is not differentiable at $(0,0)$ unless $g = 0$. *Hint:* First show that $Df(0,0)$ would have to be 0 by considering (h,k) with $k = 0$ and then with $h = 0$.

2-5. Let $f: \mathbf{R}^2 \to \mathbf{R}$ be defined by

$$f(x,y) = \begin{cases} \dfrac{x|y|}{\sqrt{x^2 + y^2}} & (x,y) \neq 0, \\ 0 & (x,y) = 0. \end{cases}$$

Show that f is a function of the kind considered in Problem 2-4, so that f is not differentiable at $(0,0)$.

2-6. Let $f: \mathbf{R}^2 \to \mathbf{R}$ be defined by $f(x,y) = \sqrt{|xy|}$. Show that f is not differentiable at $(0,0)$.

2-7. Let $f: \mathbf{R}^n \to \mathbf{R}$ be a function such that $|f(x)| \leq |x|^2$. Show that f is differentiable at 0.

2-8. Let $f: \mathbf{R} \to \mathbf{R}^2$. Prove that f is differentiable at $a \in \mathbf{R}$ if and only if f^1 and f^2 are, and that in this case

$$f'(a) = \begin{pmatrix} (f^1)'(a) \\ (f^2)'(a) \end{pmatrix}.$$

2-9. Two functions $f,g: \mathbf{R} \to \mathbf{R}$ are **equal up to nth order** at a if

$$\lim_{h \to 0} \frac{f(a + h) - g(a + h)}{h^n} = 0.$$

(a) Show that f is differentiable at a if and only if there is a function g of the form $g(x) = a_0 + a_1(x - a)$ such that f and g are equal up to first order at a.

(b) If $f'(a), \ldots, f^{(n)}(a)$ exist, show that f and the function g defined by

$$g(x) = \sum_{i=0}^{n} \frac{f^{(i)}(a)}{i!} (x - a)^i$$

are equal up to nth order at a. *Hint:* The limit

$$\lim_{x\to a}\frac{f(x)-\displaystyle\sum_{i=0}^{n-1}\frac{f^{(i)}(a)}{i!}(x-a)^i}{(x-a)^n}$$

may be evaluated by L'Hospital's rule.

BASIC THEOREMS

2-2 Theorem (Chain Rule). *If $f\colon \mathbf{R}^n \to \mathbf{R}^m$ is differentiable at a, and $g\colon \mathbf{R}^m \to \mathbf{R}^p$ is differentiable at $f(a)$, then the composition $g \circ f\colon \mathbf{R}^n \to \mathbf{R}^p$ is differentiable at a, and*

$$D(g \circ f)(a) = Dg(f(a)) \circ Df(a).$$

Remark. This equation can be written

$$(g \circ f)'(a) = g'(f(a)) \cdot f'(a).$$

If $m = n = p = 1$, we obtain the old chain rule.

Proof. Let $b = f(a)$, let $\lambda = Df(a)$, and let $\mu = Dg(f(a))$. If we define

$$\begin{aligned}
&(1)\ \ \varphi(x) = f(x) - f(a) - \lambda(x - a),\\
&(2)\ \ \psi(y) = g(y) - g(b) - \mu(y - b),\\
&(3)\ \ \rho(x) = g \circ f(x) - g \circ f(a) - \mu \circ \lambda(x - a),
\end{aligned}$$

then

$$(4)\ \ \lim_{x\to a}\frac{|\varphi(x)|}{|x-a|} = 0,$$

$$(5)\ \ \lim_{y\to b}\frac{|\psi(y)|}{|y-b|} = 0,$$

and we must show that

$$\lim_{x\to a}\frac{|\rho(x)|}{|x-a|} = 0.$$

Now

$$\begin{aligned}
\rho(x) &= g(f(x)) - g(b) - \mu(\lambda(x - a))\\
&= g(f(x)) - g(b) - \mu(f(x) - f(a) - \varphi(x)) \qquad \text{by (1)}\\
&= [g(f(x)) - g(b) - \mu(f(x) - f(a))] + \mu(\varphi(x))\\
&= \psi(f(x)) + \mu(\varphi(x)) \qquad \text{by (2).}
\end{aligned}$$

Thus we must prove

$$(6)\ \lim_{x\to a} \frac{|\psi(f(x))|}{|x-a|} = 0,$$
$$(7)\ \lim_{x\to a} \frac{|\mu(\varphi(x))|}{|x-a|} = 0.$$

Equation (7) follows easily from (4) and Problem 1-10. If $\varepsilon > 0$ it follows from (5) that for some $\delta > 0$ we have

$$|\psi(f(x))| < \varepsilon|f(x) - b| \qquad \text{if } |f(x) - b| < \delta,$$

which is true if $|x - a| < \delta_1$, for a suitable δ_1. Then

$$\begin{aligned} |\psi(f(x))| &< \varepsilon|f(x) - b| \\ &= \varepsilon|\varphi(x) + \lambda(x - a)| \\ &\leq \varepsilon|\varphi(x)| + \varepsilon M|x - a| \end{aligned}$$

for some M, by Problem 1-10. Equation (6) now follows easily. ▌

2-3 Theorem

(1) *If $f\colon \mathbf{R}^n \to \mathbf{R}^m$ is a constant function (that is, if for some $y \in \mathbf{R}^m$ we have $f(x) = y$ for all $x \in \mathbf{R}^n$), then*

$$Df(a) = 0.$$

(2) *If $f\colon \mathbf{R}^n \to \mathbf{R}^m$ is a linear transformation, then*

$$Df(a) = f.$$

(3) *If $f\colon \mathbf{R}^n \to \mathbf{R}^m$, then f is differentiable at $a \in \mathbf{R}^n$ if and only if each f^i is, and*

$$Df(a) = (Df^1(a), \ldots, Df^m(a)).$$

Thus $f'(a)$ is the $m \times n$ matrix whose ith *row is $(f^i)'(a)$.*

(4) *If $s\colon \mathbf{R}^2 \to \mathbf{R}$ is defined by $s(x,y) = x + y$, then*

$$Ds(a,b) = s.$$

(5) *If $p\colon \mathbf{R}^2 \to \mathbf{R}$ is defined by $p(x,y) = x \cdot y$, then*

$$Dp(a,b)(x,y) = bx + ay.$$

Thus $p'(a,b) = (b,a)$.

Proof

(1) $\lim\limits_{h\to 0} \dfrac{|f(a+h)-f(a)-0|}{|h|} = \lim\limits_{h\to 0} \dfrac{|y-y-0|}{|h|} = 0.$

(2) $\lim\limits_{h\to 0} \dfrac{|f(a+h)-f(a)-f(h)|}{|h|}$

$$= \lim_{h\to 0} \frac{|f(a)+f(h)-f(a)-f(h)|}{|h|} = 0.$$

(3) If each f^i is differentiable at a and

$$\lambda = (Df^1(a), \ldots, Df^m(a)),$$

then

$$\begin{aligned} f(a+h) - f(a) - \lambda(h) \\ = (f^1(a+h) - f^1(a) - Df^1(a)(h), \ldots, \\ f^m(a+h) + f^m(a) - Df^m(a)(h)). \end{aligned}$$

Therefore

$$\begin{aligned} \lim_{h\to 0} \frac{|f(a+h)-f(a)-\lambda(h)|}{|h|} \\ \leq \lim_{h\to 0} \sum_{i=1}^{m} \frac{|f^i(a+h)-f^i(a)-Df^i(a)(h)|}{|h|} = 0. \end{aligned}$$

If, on the other hand, f is differentiable at a, then $f^i = \pi^i \circ f$ is differentiable at a by (2) and Theorem 2-2.

(4) follows from (2).

(5) Let $\lambda(x,y) = bx + ay$. Then

$$\begin{aligned} \lim_{(h,k)\to 0} \frac{|p(a+h, b+k) - p(a,b) - \lambda(h,k)|}{|(h,k)|} \\ = \lim_{(h,k)\to 0} \frac{|hk|}{|(h,k)|}. \end{aligned}$$

Now

$$|hk| \leq \begin{cases} |h|^2 & \text{if } |k| \leq |h|, \\ |k|^2 & \text{if } |h| \leq |k|. \end{cases}$$

Hence $|hk| \leq |h|^2 + |k|^2$. Therefore

$$\frac{|hk|}{|(h,k)|} \leq \frac{h^2+k^2}{\sqrt{h^2+k^2}} = \sqrt{h^2+k^2},$$

so

$$\lim_{(h,k)\to 0} \frac{|hk|}{|(h,k)|} = 0. \quad \blacksquare$$

2-4 Corollary. *If* $f,g\colon \mathbf{R}^n \to \mathbf{R}$ *are differentiable at* a*, then*

$$\begin{aligned} D(f+g)(a) &= Df(a) + Dg(a), \\ D(f\cdot g)(a) &= g(a)Df(a) + f(a)Dg(a). \end{aligned}$$

If, moreover, $g(a) \neq 0$*, then*

$$D(f/g)(a) = \frac{g(a)Df(a) - f(a)Dg(a)}{[g(a)]^2}.$$

Proof. We will prove the first equation and leave the others to the reader. Since $f + g = s \circ (f,g)$, we have

$$\begin{aligned} D(f+g)(a) &= Ds(f(a),g(a)) \circ D(f,g)(a) \\ &= s \circ (Df(a),Dg(a)) \\ &= Df(a) + Dg(a). \quad \blacksquare \end{aligned}$$

We are now assured of the differentiability of those functions $f\colon \mathbf{R}^n \to \mathbf{R}^m$, whose component functions are obtained by addition, multiplication, division, and composition, from the functions π^i (which are linear transformations) and the functions which we can already differentiate by elementary calculus. Finding $Df(x)$ or $f'(x)$, however, may be a fairly formidable task. For example, let $f\colon \mathbf{R}^2 \to \mathbf{R}$ be defined by $f(x,y) = \sin(xy^2)$. Since $f = \sin \circ (\pi^1 \cdot [\pi^2]^2)$, we have

$$\begin{aligned} f'(a,b) &= \sin'(ab^2) \cdot [b^2(\pi^1)'(a,b) + a([\pi^2]^2)'(a,b)] \\ &= \sin'(ab^2) \cdot [b^2(\pi^1)'(a,b) + 2ab(\pi^2)'(a,b)] \\ &= (\cos(ab^2)) \cdot [b^2(1,0) + 2ab(0,1)] \\ &= (b^2\cos(ab^2),\ 2ab\cos(ab^2)). \end{aligned}$$

Fortunately, we will soon discover a much simpler method of computing f'.

Problems. **2-10.** Use the theorems of this section to find f' for the following:

(a) $f(x,y,z) = x^y$.
(b) $f(x,y,z) = (x^y,z)$.
(c) $f(x,y) = \sin(x \sin y)$.
(d) $f(x,y,z) = \sin(x \sin(y \sin z))$.
(e) $f(x,y,z) = x^{y^z}$.
(f) $f(x,y,z) = x^{y+z}$.
(g) $f(x,y,z) = (x + y)^z$.
(h) $f(x,y) = \sin(xy)$.
(i) $f(x,y) = [\sin(xy)]^{\cos 3}$.
(j) $f(x,y) = (\sin(xy), \sin(x \sin y), x^y)$.

2-11. Find f' for the following (where $g: \mathbf{R} \to \mathbf{R}$ is continuous):
(a) $f(x,y) = \int_a^{x+y} g$.
(b) $f(x,y) = \int_a^{x \cdot y} g$.
(c) $f(x,y,z) = \int_{xy}^{\sin(x \sin(y \sin z))} g$.

2-12. A function $f: \mathbf{R}^n \times \mathbf{R}^m \to \mathbf{R}^p$ is **bilinear** if for $x,x_1,x_2 \in \mathbf{R}^n$, $y,y_1,y_2 \in \mathbf{R}^m$, and $a \in \mathbf{R}$ we have

$$f(ax,y) = af(x,y) = f(x,ay),$$
$$f(x_1 + x_2,y) = f(x_1,y) + f(x_2,y),$$
$$f(x,y_1 + y_2) = f(x,y_1) + f(x,y_2).$$

(a) Prove that if f is bilinear, then

$$\lim_{(h,k)\to 0} \frac{|f(h,k)|}{|(h,k)|} = 0.$$

(b) Prove that $Df(a,b)(x,y) = f(a,y) + f(x,b)$.
(c) Show that the formula for $Dp(a,b)$ in Theorem 2-3 is a special case of (b).

2-13. Define $IP: \mathbf{R}^n \times \mathbf{R}^n \to \mathbf{R}$ by $IP(x,y) = \langle x,y \rangle$.
(a) Find $D(IP)(a,b)$ and $(IP)'(a,b)$.
(b) If $f,g: \mathbf{R} \to \mathbf{R}^n$ are differentiable and $h: \mathbf{R} \to \mathbf{R}$ is defined by $h(t) = \langle f(t),g(t) \rangle$, show that

$$h'(a) = \langle f'(a)^{\mathrm{T}},g(a) \rangle + \langle f(a),g'(a)^{\mathrm{T}} \rangle.$$

(Note that $f'(a)$ is an $n \times 1$ matrix; its transpose $f'(a)^{\mathrm{T}}$ is a $1 \times n$ matrix, which we consider as a member of $\mathbf{R}^n$.)
(c) If $f: \mathbf{R} \to \mathbf{R}^n$ is differentiable and $|f(t)| = 1$ for all t, show that $\langle f'(t)^{\mathrm{T}},f(t) \rangle = 0$.
(d) Exhibit a differentiable function $f: \mathbf{R} \to \mathbf{R}$ such that the function $|f|$ defined by $|f|(t) = |f(t)|$ is not differentiable.

2-14. Let E_i, $i = 1, \ldots ,k$ be Euclidean spaces of various dimensions. A function $f: E_1 \times \cdots \times E_k \to \mathbf{R}^p$ is called **multilinear** if for each choice of $x_j \in E_j$, $j \neq i$ the function $g: E_i \to \mathbf{R}^p$ defined by $g(x) = f(x_1, \ldots ,x_{i-1},x,x_{i+1}, \ldots ,x_k)$ is a linear transformation.

(a) If f is multilinear and $i \neq j$, show that for $h = (h_1, \ldots, h_k)$, with $h_l \in E_l$, we have

$$\lim_{h \to 0} \frac{|f(a_1, \ldots, h_i, \ldots, h_j, \ldots, a_k)|}{|h|} = 0.$$

Hint: If $g(x,y) = f(a_1, \ldots, x, \ldots, y, \ldots, a_k)$, then g is bilinear.

(b) Prove that

$$Df(a_1, \ldots, a_k)(x_1, \ldots, x_k) = \sum_{i=1}^{k} f(a_1, \ldots, a_{i-1}, x_i, a_{i+1}, \ldots, a_k).$$

2-15. Regard an $n \times n$ matrix as a point in the n-fold product $\mathbf{R}^n \times \cdots \times \mathbf{R}^n$ by considering each row as a member of $\mathbf{R}^n$.

(a) Prove that $\det: \mathbf{R}^n \times \cdots \times \mathbf{R}^n \to \mathbf{R}$ is differentiable and

$$D(\det)(a_1, \ldots, a_n)(x_1, \ldots, x_n) = \sum_{i=1}^{n} \det \begin{pmatrix} a_1 \\ \cdot \\ \cdot \\ \cdot \\ x_i \\ \cdot \\ \cdot \\ \cdot \\ a_n \end{pmatrix}.$$

(b) If $a_{ij}: \mathbf{R} \to \mathbf{R}$ are differentiable and $f(t) = \det(a_{ij}(t))$, show that

$$f'(t) = \sum_{j=1}^{n} \det \begin{pmatrix} a_{11}(t), & \ldots, & a_{1n}(t) \\ \cdot & & \cdot \\ \cdot & & \cdot \\ \cdot & & \cdot \\ a_{j1}'(t), & \ldots, & a_{jn}'(t) \\ \cdot & & \cdot \\ \cdot & & \cdot \\ \cdot & & \cdot \\ a_{n1}(t), & \ldots, & a_{nn}(t) \end{pmatrix}.$$

(c) If $\det(a_{ij}(t)) \neq 0$ for all t and $b_1, \ldots, b_n: \mathbf{R} \to \mathbf{R}$ are differentiable, let $s_1, \ldots, s_n: \mathbf{R} \to \mathbf{R}$ be the functions such that $s_1(t), \ldots, s_n(t)$ are the solutions of the equations

$$\sum_{j=1}^{n} a_{ji}(t) s_j(t) = b_i(t) \qquad i = 1, \ldots, n.$$

Show that s_i is differentiable and find $s_i'(t)$.

2-16. Suppose $f\colon \mathbf{R}^n \to \mathbf{R}^n$ is differentiable and has a differentiable inverse $f^{-1}\colon \mathbf{R}^n \to \mathbf{R}^n$. Show that $(f^{-1})'(a) = [f'(f^{-1}(a))]^{-1}$. *Hint:* $f \circ f^{-1}(x) = x$.

PARTIAL DERIVATIVES

We begin the attack on the problem of finding derivatives "one variable at a time." If $f\colon \mathbf{R}^n \to \mathbf{R}$ and $a \in \mathbf{R}^n$, the limit

$$\lim_{h\to 0} \frac{f(a^1, \ldots, a^i + h, \ldots, a^n) - f(a^1, \ldots, a^n)}{h},$$

if it exists, is denoted $D_i f(a)$, and called the ith **partial derivative** of f at a. It is important to note that $D_i f(a)$ is the ordinary derivative of a certain function; in fact, if $g(x) = f(a^1, \ldots, x, \ldots, a^n)$, then $D_i f(a) = g'(a^i)$. This means that $D_i f(a)$ is the slope of the tangent line at $(a, f(a))$ to the curve obtained by intersecting the graph of f with the plane $x^j = a^j$, $j \neq i$ (Figure 2-1). It also means that computation of $D_i f(a)$ is a problem we can already solve. If $f(x^1, \ldots, x^n)$ is

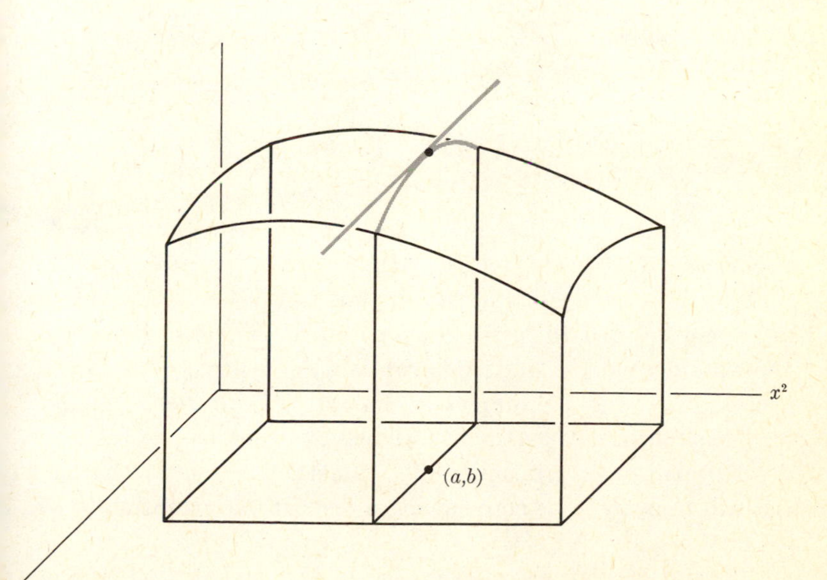

FIGURE 2-1

given by some formula involving $x^1, \ldots, x^n$, then we find $D_if(x^1, \ldots, x^n)$ by differentiating the function whose value at x^i is given by the formula when all x^j, for $j \neq i$, are thought of as constants. For example, if $f(x,y) = \sin(xy^2)$, then $D_1f(x,y) = y^2 \cos(xy^2)$ and $D_2f(x,y) = 2xy\cos(xy^2)$. If, instead, $f(x,y) = x^y$, then $D_1f(x,y) = yx^{y-1}$ and $D_2f(x,y) = x^y \log x$.

With a little practice (e.g., the problems at the end of this section) you should acquire as great a facility for computing D_if as you already have for computing ordinary derivatives.

If $D_if(x)$ exists for all $x \in \mathbf{R}^n$, we obtain a function $D_if\colon \mathbf{R}^n \to \mathbf{R}$. The jth partial derivative of this function at x, that is, $D_j(D_if)(x)$, is often denoted $D_{i,j}f(x)$. Note that this notation reverses the order of i and j. As a matter of fact, the order is usually irrelevant, since most functions (an exception is given in the problems) satisfy $D_{i,j}f = D_{j,i}f$. There are various delicate theorems ensuring this equality; the following theorem is quite adequate. We state it here but postpone the proof until later (Problem 3-28).

2-5 Theorem. *If $D_{i,j}f$ and $D_{j,i}f$ are continuous in an open set containing a, then*

$$D_{i,j}f(a) = D_{j,i}f(a).$$

The function $D_{i,j}f$ is called a **second-order (mixed) partial derivative** of f. Higher-order (mixed) partial derivatives are defined in the obvious way. Clearly Theorem 2-5 can be used to prove the equality of higher-order mixed partial derivatives under appropriate conditions. The order of $i_1, \ldots, i_k$ is completely immaterial in $D_{i_1, \ldots, i_k}f$ if f has continuous partial derivatives of all orders. A function with this property is called a C^∞ function. In later chapters it will frequently be convenient to restrict our attention to C^∞ functions.

Partial derivatives will be used in the next section to find derivatives. They also have another important use—finding maxima and minima of functions.

2-6 Theorem. *Let $A \subset \mathbf{R}^n$. If the maximum (or minimum) of $f: A \to \mathbf{R}$ occurs at a point a in the interior of A and $D_i f(a)$ exists, then $D_i f(a) = 0$.*

Proof. Let $g_i(x) = f(a^1, \ldots, x, \ldots, a^n)$. Clearly g_i has a maximum (or minimum) at a^i, and g_i is defined in an open interval containing a^i. Hence $0 = g_i'(a^i) = D_i f(a)$. ∎

The reader is reminded that the converse of Theorem 2-6 is false even if $n = 1$ (if $f: \mathbf{R} \to \mathbf{R}$ is defined by $f(x) = x^3$, then $f'(0) = 0$, but 0 is not even a local maximum or minimum). If $n > 1$, the converse of Theorem 2-6 may fail to be true in a rather spectacular way. Suppose, for example, that $f: \mathbf{R}^2 \to \mathbf{R}$ is defined by $f(x,y) = x^2 - y^2$ (Figure 2-2). Then $D_1 f(0,0) = 0$ because g_1 has a minimum at 0, while $D_2 f(0,0) = 0$ because g_2 has a maximum at 0. Clearly (0,0) is neither a relative maximum nor a relative minimum.

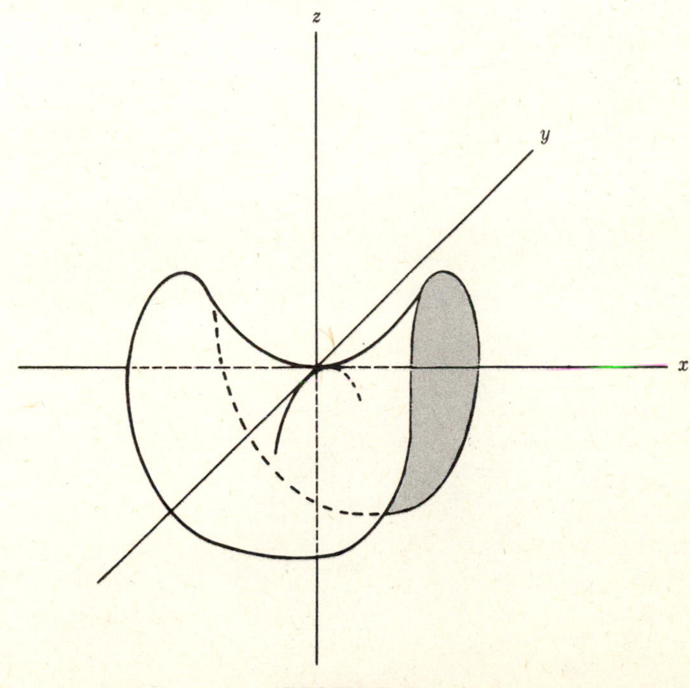

FIGURE 2-2

If Theorem 2-6 is used to find the maximum or minimum of f on A, the values of f at boundary points must be examined separately—a formidable task, since the boundary of A may be all of A! Problem 2-27 indicates one way of doing this, and Problem 5-16 states a superior method which can often be used.

Problems. **2-17.** Find the partial derivatives of the following functions:

(a) $f(x,y,z) = x^y$.
(b) $f(x,y,z) = z$.
(c) $f(x,y) = \sin(x \sin y)$.
(d) $f(x,y,z) = \sin(x \sin(y \sin z))$.
(e) $f(x,y,z) = x^{y^z}$.
(f) $f(x,y,z) = x^{y+z}$.
(g) $f(x,y,z) = (x + y)^z$.
(h) $f(x,y) = \sin(xy)$.
(i) $f(x,y) = [\sin(xy)]^{\cos 3}$.

2-18. Find the partial derivatives of the following functions (where $g\colon \mathbf{R} \to \mathbf{R}$ is continuous):

(a) $f(x,y) = \int_a^{x+y} g$.
(b) $f(x,y) = \int_y^x g$.
(c) $f(x,y) = \int_a^{xy} g$.
(d) $f(x,y) = \int_a^{\left(\int_b^y g\right)} g$.

2-19. If $f(x,y) = x^{x^{x^{x^y}}} + (\log x)(\arctan(\arctan(\arctan(\sin(\cos xy) - \log(x + y)))))$ find $D_2f(1,y)$. *Hint:* There is an easy way to do this.

2-20. Find the partial derivatives of f in terms of the derivatives of g and h if

(a) $f(x,y) = g(x)h(y)$.
(b) $f(x,y) = g(x)^{h(y)}$.
(c) $f(x,y) = g(x)$.
(d) $f(x,y) = g(y)$.
(e) $f(x,y) = g(x + y)$.

2-21.* Let $g_1,g_2\colon \mathbf{R}^2 \to \mathbf{R}$ be continuous. Define $f\colon \mathbf{R}^2 \to \mathbf{R}$ by

$$f(x,y) = \int_0^x g_1(t,0)dt + \int_0^y g_2(x,t)dt.$$

(a) Show that $D_2f(x,y) = g_2(x,y)$.
(b) How should f be defined so that $D_1f(x,y) = g_1(x,y)$?
(c) Find a function $f\colon \mathbf{R}^2 \to \mathbf{R}$ such that $D_1f(x,y) = x$ and $D_2f(x,y) = y$. Find one such that $D_1f(x,y) = y$ and $D_2f(x,y) = x$.

2-22.* If $f: \mathbf{R}^2 \to \mathbf{R}$ and $D_2 f = 0$, show that f is independent of the second variable. If $D_1 f = D_2 f = 0$, show that f is constant.

2-23.* Let $A = \{(x,y) \in \mathbf{R}^2: x < 0, \text{ or } x \geq 0 \text{ and } y \neq 0\}$.

(a) If $f: A \to \mathbf{R}$ and $D_1 f = D_2 f = 0$, show that f is constant. *Hint:* Note that any two points in A can be connected by a sequence of lines each parallel to one of the axes.

(b) Find a function $f: A \to \mathbf{R}$ such that $D_2 f = 0$ but f is not independent of the second variable.

2-24. Define $f: \mathbf{R}^2 \to \mathbf{R}$ by

$$f(x,y) = \begin{cases} xy\dfrac{x^2 - y^2}{x^2 + y^2} & (x,y) \neq 0, \\ 0 & (x,y) = 0. \end{cases}$$

(a) Show that $D_2 f(x,0) = x$ for all x and $D_1 f(0,y) = -y$ for all y.

(b) Show that $D_{1,2} f(0,0) \neq D_{2,1} f(0,0)$.

2-25.* Define $f: \mathbf{R} \to \mathbf{R}$ by

$$f(x) = \begin{cases} e^{-x^{-2}} & x \neq 0, \\ 0 & x = 0. \end{cases}$$

Show that f is a C^∞ function, and $f^{(i)}(0) = 0$ for all i. *Hint:* The limit $f'(0) = \lim\limits_{h \to 0} \dfrac{e^{-h^{-2}}}{h} = \lim\limits_{h \to 0} \dfrac{1/h}{e^{h^{-2}}}$ can be evaluated by L'Hospital's rule. It is easy enough to find $f'(x)$ for $x \neq 0$, and $f''(0) = \lim\limits_{h \to 0} f'(h)/h$ can then be found by L'Hospital's rule.

2-26.* Let $\quad f(x) = \begin{cases} e^{-(x-1)^{-2}} \cdot e^{-(x+1)^{-2}} & x \in (-1,1), \\ 0 & x \notin (-1,1). \end{cases}$

(a) Show that $f: \mathbf{R} \to \mathbf{R}$ is a C^∞ function which is positive on $(-1,1)$ and 0 elsewhere.

(b) Show that there is a C^∞ function $g: \mathbf{R} \to [0,1]$ such that $g(x) = 0$ for $x \leq 0$ and $g(x) = 1$ for $x \geq \varepsilon$. *Hint:* If f is a C^∞ function which is positive on $(0,\varepsilon)$ and 0 elsewhere, let $g(x) = \int_0^x f / \int_0^\varepsilon f$.

(c) If $a \in \mathbf{R}^n$, define $g: \mathbf{R}^n \to \mathbf{R}$ by

$$g(x) = f([x^1 - a^1]/\varepsilon) \cdot \ldots \cdot f([x^n - a^n]/\varepsilon).$$

Show that g is a C^∞ function which is positive on

$$(a^1 - \varepsilon, a^1 + \varepsilon) \times \cdots \times (a^n - \varepsilon, a^n + \varepsilon)$$

and zero elsewhere.

(d) If $A \subset \mathbf{R}^n$ is open and $C \subset A$ is compact, show that there is a non-negative C^∞ function $f: A \to \mathbf{R}$ such that $f(x) > 0$ for $x \in C$ and $f = 0$ outside of some closed set contained in A.

(e) Show that we can choose such an f so that $f: A \to [0,1]$ and $f(x) = 1$ for $x \in C$. *Hint:* If the function f of (d) satisfies $f(x) \geq \varepsilon$ for $x \in C$, consider $g \circ f$, where g is the function of (b).

2-27. Define $g, h: \{x \in \mathbf{R}^2: |x| \le 1\} \to \mathbf{R}^3$ by

$$g(x,y) = (x,y,\sqrt{1-x^2-y^2}),$$
$$h(x,y) = (x,y,-\sqrt{1-x^2-y^2}).$$

Show that the maximum of f on $\{x \in \mathbf{R}^3: |x| = 1\}$ is either the maximum of $f \circ g$ or the maximum of $f \circ h$ on $\{x \in \mathbf{R}^2: |x| \le 1\}$.

DERIVATIVES

The reader who has compared Problems 2-10 and 2-17 has probably already guessed the following.

2-7 *Theorem.* *If $f: \mathbf{R}^n \to \mathbf{R}^m$ is differentiable at a, then $D_jf^i(a)$ exists for $1 \le i \le m$, $1 \le j \le n$ and $f'(a)$ is the $m \times n$ matrix $(D_jf^i(a))$.*

Proof. Suppose first that $m = 1$, so that $f: \mathbf{R}^n \to \mathbf{R}$. Define $h: \mathbf{R} \to \mathbf{R}^n$ by $h(x) = (a^1, \ldots, x, \ldots, a^n)$, with x in the jth place. Then $D_jf(a) = (f \circ h)'(a^j)$. Hence, by Theorem 2-2,

$$(f \circ h)'(a^j) = f'(a) \cdot h'(a^j)$$
$$= f'(a) \cdot \begin{pmatrix} 0 \\ \cdot \\ \cdot \\ \cdot \\ 1 \\ \cdot \\ \cdot \\ \cdot \\ 0 \end{pmatrix} \leftarrow j\text{th place.}$$

Since $(f \circ h)'(a^j)$ has the single entry $D_jf(a)$, this shows that $D_jf(a)$ exists and is the jth entry of the $1 \times n$ matrix $f'(a)$.

The theorem now follows for arbitrary m since, by Theorem 2-3, each f^i is differentiable and the ith row of $f'(a)$ is $(f^i)'(a)$. ∎

There are several examples in the problems to show that the converse of Theorem 2-7 is false. It is true, however, if one hypothesis is added.

2-8 Theorem. *If $f\colon \mathbf{R}^n \to \mathbf{R}^m$, then $Df(a)$ exists if all $D_jf^i(x)$ exist in an open set containing a and if each function D_jf^i is continuous at a.*
(Such a function f is called **continuously differentiable** at a.)

Proof. As in the proof of Theorem 2-7, it suffices to consider the case $m = 1$, so that $f\colon \mathbf{R}^n \to \mathbf{R}$. Then

$$\begin{aligned} f(a+h) - f(a) = {} & f(a^1 + h^1, a^2, \ldots, a^n) - f(a^1, \ldots, a^n) \\ & + f(a^1 + h^1, a^2 + h^2, a^3, \ldots, a^n) \\ & \qquad - f(a^1 + h^1, a^2, \ldots, a^n) \\ & + \cdots \\ & + f(a^1 + h^1, \ldots, a^n + h^n) \\ & \qquad - f(a^1 + h^1, \ldots, a^{n-1} + h^{n-1}, a^n). \end{aligned}$$

Recall that D_1f is the derivative of the function g defined by $g(x) = f(x, a^2, \ldots, a^n)$. Applying the mean-value theorem to g we obtain

$$\begin{aligned} f(a^1 + h^1, a^2, \ldots, a^n) - f(a^1, \ldots, a^n) \\ = h^1 \cdot D_1f(b_1, a^2, \ldots, a^n) \end{aligned}$$

for some b_1 between a^1 and $a^1 + h^1$. Similarly the ith term in the sum equals

$$h^i \cdot D_if(a^1 + h^1, \ldots, a^{i-1} + h^{i-1}, b_i, \ldots, a^n) = h^i D_if(c_i),$$

for some c_i. Then

$$\begin{aligned} \lim_{h \to 0} \frac{\left| f(a+h) - f(a) - \sum_{i=1}^{n} D_if(a) \cdot h^i \right|}{|h|} & = \lim_{h \to 0} \frac{\left| \sum_{i=1}^{n} [D_if(c_i) - D_if(a)] \cdot h^i \right|}{|h|} \\ & \le \lim_{h \to 0} \sum_{i=1}^{n} |D_if(c_i) - D_if(a)| \cdot \frac{|h^i|}{|h|} \\ & \le \lim_{h \to 0} \sum_{i=1}^{n} |D_if(c_i) - D_if(a)| \\ & = 0, \end{aligned}$$

since D_if is continuous at a. ▌

Although the chain rule was used in the proof of Theorem 2-7, it could easily have been eliminated. With Theorem 2-8 to provide differentiable functions, and Theorem 2-7 to provide their derivatives, the chain rule may therefore seem almost superfluous. However, it has an extremely important corollary concerning partial derivatives.

2-9 Theorem. *Let $g_1, \ldots, g_m: \mathbf{R}^n \to \mathbf{R}$ be continuously differentiable at a, and let $f: \mathbf{R}^m \to \mathbf{R}$ be differentiable at $(g_1(a), \ldots, g_m(a))$. Define the function $F: \mathbf{R}^n \to \mathbf{R}$ by $F(x) = f(g_1(x), \ldots, g_m(x))$. Then*

$$D_iF(a) = \sum_{j=1}^{m} D_jf(g_1(a), \ldots, g_m(a)) \cdot D_ig_j(a).$$

Proof. The function F is just the composition $f \circ g$, where $g = (g_1, \ldots, g_m)$. Since g_i is continuously differentiable at a, it follows from Theorem 2-8 that g is differentiable at a. Hence by Theorem 2-2,

$$F'(a) = f'(g(a)) \cdot g'(a) =$$

$$(D_1f(g(a)), \ldots, D_mf(g(a))) \cdot \begin{pmatrix} D_1g_1(a), & \cdots & ,D_ng_1(a) \\ \vdots & & \vdots \\ D_1g_m(a), & \ldots & ,D_ng_m(a) \end{pmatrix}$$

But $D_iF(a)$ is the ith entry of the left side of this equation, while $\Sigma_{j=1}^m D_jf(g_1(a), \ldots, g_m(a)) \cdot D_ig_j(a)$ is the ith entry of the right side. ▌

Theorem 2-9 is often called the *chain rule*, but is weaker than Theorem 2-2 since g could be differentiable without g_i being continuously differentiable (see Problem 2-32). Most computations requiring Theorem 2-9 are fairly straightforward. A slight subtlety is required for the function $F: \mathbf{R}^2 \to \mathbf{R}$ defined by

$$F(x,y) = f(g(x,y),h(x),k(y))$$

where $h,k\colon \mathbf{R} \to \mathbf{R}$. In order to apply Theorem 2-9 define $\bar{h},\bar{k}\colon \mathbf{R}^2 \to \mathbf{R}$ by

$$\bar{h}(x,y) = h(x) \qquad \bar{k}(x,y) = k(y).$$

Then

$$\begin{aligned} D_1\bar{h}(x,y) &= h'(x) & D_2\bar{h}(x,y) &= 0, \\ D_1\bar{k}(x,y) &= 0 & D_2\bar{k}(x,y) &= k'(y), \end{aligned}$$

and we can write

$$F(x,y) = f(g(x,y),\bar{h}(x,y),\bar{k}(x,y)).$$

Letting $a = (g(x,y),h(x),k(y))$, we obtain

$$\begin{aligned} D_1F(x,y) &= D_1f(a) \cdot D_1g(x,y) + D_2f(a) \cdot h'(x), \\ D_2F(x,y) &= D_1f(a) \cdot D_2g(x,y) + D_3f(a) \cdot k'(y). \end{aligned}$$

It should, of course, be unnecessary for you to actually write down the functions $\bar{h}$ and $\bar{k}$.

Problems. **2-28.** Find expressions for the partial derivatives of the following functions:

(a) $F(x,y) = f(g(x)k(y),\ g(x) + h(y))$.
(b) $F(x,y,z) = f(g(x + y),\ h(y + z))$.
(c) $F(x,y,z) = f(x^y,y^z,z^x)$.
(d) $F(x,y) = f(x,g(x),h(x,y))$.

2-29. Let $f\colon \mathbf{R}^n \to \mathbf{R}$. For $x \in \mathbf{R}^n$, the limit

$$\lim_{t\to 0} \frac{f(a + tx) - f(a)}{t},$$

if it exists, is denoted $D_xf(a)$, and called the **directional derivative** of f at a, in the direction x.

(a) Show that $D_{e_i}f(a) = D_if(a)$.
(b) Show that $D_{tx}f(a) = tD_xf(a)$.
(c) If f is differentiable at a, show that $D_xf(a) = Df(a)(x)$ and therefore $D_{x+y}f(a) = D_xf(a) + D_yf(a)$.

2-30. Let f be defined as in Problem 2-4. Show that $D_xf(0,0)$ exists for all x, but if $g \neq 0$, then $D_{x+y}f(0,0) = D_xf(0,0) + D_yf(0,0)$ is not true for all x and y.

2-31. Let $f\colon \mathbf{R}^2 \to \mathbf{R}$ be defined as in Problem 1-26. Show that $D_xf(0,0)$ exists for all x, although f is not even continuous at $(0,0)$.

2-32. (a) Let $f\colon \mathbf{R} \to \mathbf{R}$ be defined by

$$f(x) = \begin{cases} x^2 \sin \dfrac{1}{x} & x \neq 0, \\ 0 & x = 0. \end{cases}$$

Show that f is differentiable at 0 but f' is not continuous at 0.
(b) Let $f\colon \mathbf{R}^2 \to \mathbf{R}$ be defined by

$$f(x,y) = \begin{cases} (x^2 + y^2)\sin \dfrac{1}{\sqrt{x^2 + y^2}} & (x,y) \neq 0, \\ 0 & (x,y) = 0. \end{cases}$$

Show that f is differentiable at (0,0) but D_if is not continuous at (0,0).

2-33. Show that the continuity of D_1f^j at a may be eliminated from the hypothesis of Theorem 2-8.

2-34. A function $f\colon \mathbf{R}^n \to \mathbf{R}$ is **homogeneous** of degree m if $f(tx) = t^mf(x)$ for all x. If f is also differentiable, show that

$$\sum_{i=1}^{n} x^iD_if(x) = mf(x).$$

Hint: If $g(t) = f(tx)$, find $g'(1)$.

2-35. If $f\colon \mathbf{R}^n \to \mathbf{R}$ is differentiable and $f(0) = 0$, prove that there exist $g_i\colon \mathbf{R}^n \to \mathbf{R}$ such that

$$f(x) = \sum_{i=1}^{n} x^ig_i(x).$$

Hint: If $h_x(t) = f(tx)$, then $f(x) = \int_0^1 h_x'(t)dt$.

INVERSE FUNCTIONS

Suppose that $f\colon \mathbf{R} \to \mathbf{R}$ is continuously differentiable in an open set containing a and $f'(a) \neq 0$. If $f'(a) > 0$, there is an open interval V containing a such that $f'(x) > 0$ for $x \in V$, and a similar statement holds if $f'(a) < 0$. Thus f is increasing (or decreasing) on V, and is therefore 1-1 with an inverse function f^{-1} defined on some open interval W containing $f(a)$. Moreover it is not hard to show that f^{-1} is differentiable, and for $y \in W$ that

$$(f^{-1})'(y) = \frac{1}{f'(f^{-1}(y))}.$$

An analogous discussion in higher dimensions is much more involved, but the result (Theorem 2-11) is very important. We begin with a simple lemma.

2-10 Lemma. *Let $A \subset \mathbf{R}^n$ be a rectangle and let $f\colon A \to \mathbf{R}^n$ be continuously differentiable. If there is a number M such that $|D_j f^i(x)| \leq M$ for all x in the interior of A, then*

$$|f(x) - f(y)| \leq n^2 M |x - y|$$

for all $x,y \in A$.

Proof. We have

$$f^i(y) - f^i(x) = \sum_{j=1}^{n} [f^i(y^1, \ldots, y^j, x^{j+1}, \ldots, x^n) - f^i(y^1, \ldots, y^{j-1}, x^j, \ldots, x^n)].$$

Applying the mean-value theorem we obtain

$$f^i(y^1, \ldots, y^j, x^{j+1}, \ldots, x^n) - f^i(y^1, \ldots, y^{j-1}, x^j, \ldots, x^n) = (y^j - x^j) \cdot D_j f^i(z_{ij})$$

for some z_{ij}. The expression on the right has absolute value less than or equal to $M \cdot |y^j - x^j|$. Thus

$$|f^i(y) - f^i(x)| \leq \sum_{j=1}^{n} |y^j - x^j| \cdot M \leq nM|y - x|$$

since each $|y^j - x^j| \leq |y - x|$. Finally

$$|f(y) - f(x)| \leq \sum_{i=1}^{n} |f^i(y) - f^i(x)| \leq n^2 M \cdot |y - x|. \blacksquare$$

2-11 Theorem (*Inverse Function Theorem*). *Suppose that $f\colon \mathbf{R}^n \to \mathbf{R}^n$ is continuously differentiable in an open set containing a, and $\det f'(a) \neq 0$. Then there is an open set V containing a and an open set W containing $f(a)$ such that $f\colon V \to W$ has a continuous inverse $f^{-1}\colon W \to V$ which is differentiable and for all $y \in W$ satisfies*

$$(f^{-1})'(y) = [f'(f^{-1}(y))]^{-1}.$$

Proof. Let λ be the linear transformation $Df(a)$. Then λ is non-singular, since $\det f'(a) \neq 0$. Now $D(\lambda^{-1} \circ f)(a) = D(\lambda^{-1})(f(a)) \circ Df(a) = \lambda^{-1} \circ Df(a)$ is the identity linear transformation. If the theorem is true for $\lambda^{-1} \circ f$, it is clearly true for f. Therefore we may assume at the outset that λ is the identity. Thus whenever $f(a + h) = f(a)$, we have

$$\frac{|f(a + h) - f(a) - \lambda(h)|}{|h|} = \frac{|h|}{|h|} = 1.$$

But

$$\lim_{h \to 0} \frac{|f(a + h) - f(a) - \lambda(h)|}{|h|} = 0.$$

This means that we cannot have $f(x) = f(a)$ for x arbitrarily close to, but unequal to, a. Therefore there is a closed rectangle U containing a in its interior such that

1. $f(x) \neq f(a)$ if $x \in U$ and $x \neq a$.

Since f is continuously differentiable in an open set containing a, we can also assume that

2. $\det f'(x) \neq 0$ for $x \in U$.
3. $|D_jf^i(x) - D_jf^i(a)| < 1/2n^2$ for all i, j, and $x \in U$.

Note that (3) and Lemma 2-10 applied to $g(x) = f(x) - x$ imply for $x_1,x_2 \in U$ that

$$|f(x_1) - x_1 - (f(x_2) - x_2)| \leq \tfrac{1}{2}|x_1 - x_2|.$$

Since

$$\begin{aligned} |x_1 - x_2| - |f(x_1) - f(x_2)| &\leq |f(x_1) - x_1 - (f(x_2) - x_2)| \\ &\leq \tfrac{1}{2}|x_1 - x_2|, \end{aligned}$$

we obtain

4. $|x_1 - x_2| \leq 2|f(x_1) - f(x_2)|$ for $x_1,x_2 \in U$.

Now $f(\text{boundary } U)$ is a compact set which, by (1), does not contain $f(a)$ (Figure 2-3). Therefore there is a number $d > 0$ such that $|f(a) - f(x)| \geq d$ for $x \in \text{boundary } U$. Let $W = \{y\colon |y - f(a)| < d/2\}$. If $y \in W$ and $x \in \text{boundary } U$, then

5. $|y - f(a)| < |y - f(x)|$.

We will show that for any $y \in W$ there is a unique x in interior U such that $f(x) = y$. To prove this consider the

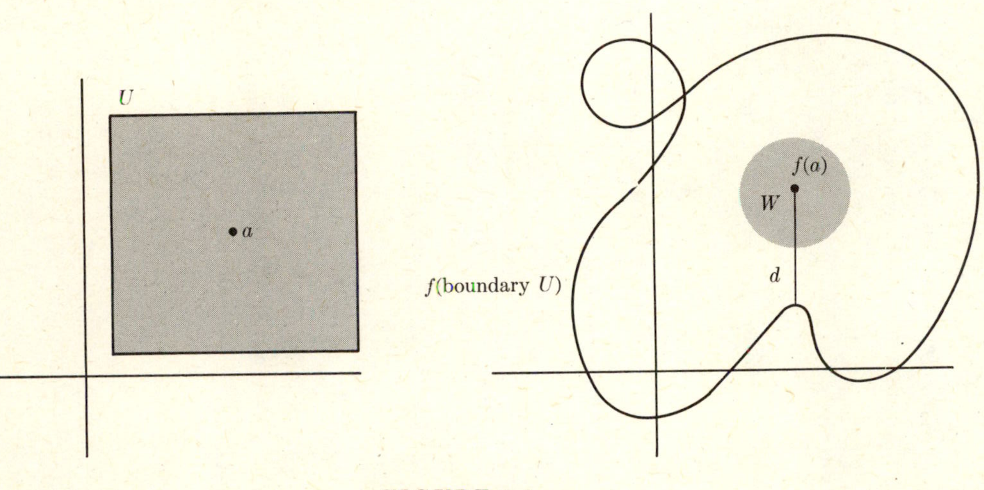

FIGURE 2-3

function $g\colon U \to \mathbf{R}$ defined by

$$g(x) = |y - f(x)|^2 = \sum_{i=1}^{n} (y^i - f^i(x))^2.$$

This function is continuous and therefore has a minimum on U. If $x \in$ boundary U, then, by (5), we have $g(a) < g(x)$. Therefore the minimum of g does *not* occur on the boundary of U. By Theorem 2-6 there is a point $x \in$ interior U such that $D_j g(x) = 0$ for all j, that is

$$\sum_{i=1}^{n} 2(y^i - f^i(x)) \cdot D_j f^i(x) = 0 \qquad \text{for all } j.$$

By (2) the matrix $(D_j f^i(x))$ has non-zero determinant. Therefore we must have $y^i - f^i(x) = 0$ for all i, that is $y = f(x)$. This proves the existence of x. Uniqueness follows immediately from (4).

If $V = (\text{interior } U) \cap f^{-1}(W)$, we have shown that the function $f\colon V \to W$ has an inverse $f^{-1}\colon W \to V$. We can rewrite (4) as

6. $|f^{-1}(y_1) - f^{-1}(y_2)| \le 2|y_1 - y_2| \qquad \text{for } y_1, y_2 \in W.$

This shows that f^{-1} is continuous.

Only the proof that f^{-1} is differentiable remains. Let $\mu = Df(x)$. We will show that f^{-1} is differentiable at $y = f(x)$ with derivative μ^{-1}. As in the proof of Theorem 2-2, for $x_1 \in V$, we have

$$f(x_1) = f(x) + \mu(x_1 - x) + \varphi(x_1 - x),$$

where

$$\lim_{x_1 \to x} \frac{|\varphi(x_1 - x)|}{|x_1 - x|} = 0.$$

Therefore

$$\mu^{-1}(f(x_1) - f(x)) = x_1 - x + \mu^{-1}(\varphi(x_1 - x)).$$

Since every $y_1 \in W$ is of the form $f(x_1)$ for some $x_1 \in V$, this can be written

$$f^{-1}(y_1) = f^{-1}(y) + \mu^{-1}(y_1 - y) - \mu^{-1}(\varphi(f^{-1}(y_1) - f^{-1}(y))),$$

and it therefore suffices to show that

$$\lim_{y_1 \to y} \frac{|\mu^{-1}(\varphi(f^{-1}(y_1) - f^{-1}(y)))|}{|y_1 - y|} = 0.$$

Therefore (Problem 1-10) it suffices to show that

$$\lim_{y_1 \to y} \frac{|\varphi(f^{-1}(y_1) - f^{-1}(y))|}{|y_1 - y|} = 0.$$

Now

$$\frac{|\varphi(f^{-1}(y_1) - f^{-1}(y))|}{|y_1 - y|} = \frac{|\varphi(f^{-1}(y_1) - f^{-1}(y))|}{|f^{-1}(y_1) - f^{-1}(y)|} \cdot \frac{|f^{-1}(y_1) - f^{-1}(y)|}{|y_1 - y|}.$$

Since f^{-1} is continuous, $f^{-1}(y_1) \to f^{-1}(y)$ as $y_1 \to y$. Therefore the first factor approaches 0. Since, by (6), the second factor is less than 2, the product also approaches 0. ▌

It should be noted that an inverse function f^{-1} may exist even if $\det f'(a) = 0$. For example, if $f\colon \mathbf{R} \to \mathbf{R}$ is defined by $f(x) = x^3$, then $f'(0) = 0$ but f has the inverse function $f^{-1}(x) = \sqrt[3]{x}$. One thing is certain however: if $\det f'(a) = 0$, then f^{-1} cannot be differentiable at $f(a)$. To prove this note that $f \circ f^{-1}(x) = x$. If f^{-1} were differentiable at $f(a)$, the chain rule would give $f'(a) \cdot (f^{-1})'(f(a)) = I$, and consequently $\det f'(a) \cdot \det(f^{-1})'(f(a)) = 1$, contradicting $\det f'(a) = 0$.

Problems. **2-36.*** Let $A \subset \mathbf{R}^n$ be an open set and $f\colon A \to \mathbf{R}^n$ a continuously differentiable 1-1 function such that $\det f'(x) \neq 0$ for all x. Show that $f(A)$ is an open set and $f^{-1}\colon f(A) \to A$ is differentiable. Show also that $f(B)$ is open for any open set $B \subset A$.

2-37. (a) Let $f\colon \mathbf{R}^2 \to \mathbf{R}$ be a continuously differentiable function. Show that f is *not* 1-1. *Hint:* If, for example, $D_1 f(x,y) \neq 0$ for all (x,y) in some open set A, consider $g\colon A \to \mathbf{R}^2$ defined by $g(x,y) = (f(x,y),y)$.

(b) Generalize this result to the case of a continuously differentiable function $f\colon \mathbf{R}^n \to \mathbf{R}^m$ with $m < n$.

2-38. (a) If $f\colon \mathbf{R} \to \mathbf{R}$ satisfies $f'(a) \neq 0$ for all $a \in \mathbf{R}$, show that f is 1-1 (on all of $\mathbf{R}$).

(b) Define $f\colon \mathbf{R}^2 \to \mathbf{R}^2$ by $f(x,y) = (e^x \cos y, e^x \sin y)$. Show that $\det f'(x,y) \neq 0$ for all (x,y) but f is not 1-1.

2-39. Use the function $f\colon \mathbf{R} \to \mathbf{R}$ defined by

$$f(x) = \begin{cases} \dfrac{x}{2} + x^2 \sin \dfrac{1}{x} & x \neq 0, \\ 0 & x = 0, \end{cases}$$

to show that continuity of the derivative cannot be eliminated from the hypothesis of Theorem 2-11.

IMPLICIT FUNCTIONS

Consider the function $f\colon \mathbf{R}^2 \to \mathbf{R}$ defined by $f(x,y) = x^2 + y^2 - 1$. If we choose (a,b) with $f(a,b) = 0$ and $a \neq 1, -1$, there are (Figure 2-4) open intervals A containing a and B containing b with the following property: if $x \in A$, there is a unique $y \in B$ with $f(x,y) = 0$. We can therefore define

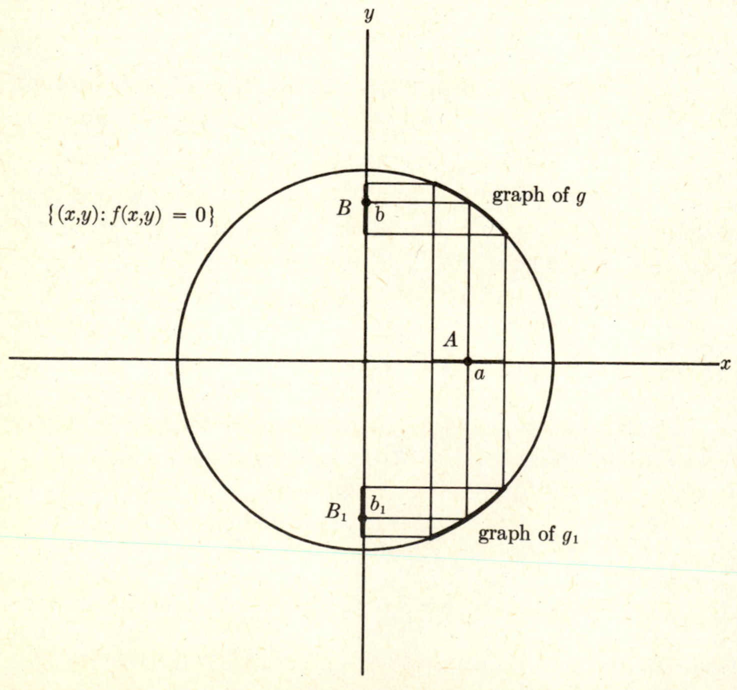

FIGURE 2-4

a function $g\colon A \to \mathbf{R}$ by the condition $g(x) \in B$ and $f(x,g(x)) = 0$ (if $b > 0$, as indicated in Figure 2-4, then $g(x) = \sqrt{1 - x^2}$). For the function f we are considering there is another number b_1 such that $f(a,b_1) = 0$. There will also be an interval B_1 containing b_1 such that, when $x \in A$, we have $f(x,g_1(x)) = 0$ for a unique $g_1(x) \in B_1$ (here $g_1(x) = -\sqrt{1 - x^2}$). Both g and g_1 are differentiable. These functions are said to be defined **implicitly** by the equation $f(x,y) = 0$.

If we choose $a = 1$ or -1 it is impossible to find any such function g defined in an open interval containing a. We would like a simple criterion for deciding when, in general, such a function can be found. More generally we may ask the following: If $f\colon \mathbf{R}^n \times \mathbf{R} \to \mathbf{R}$ and $f(a^1, \ldots ,a^n,b) = 0$, when can we find, for each $(x^1, \ldots ,x^n)$ near $(a^1, \ldots ,a^n)$, a unique y near b such that $f(x^1, \ldots ,x^n,y) = 0$? Even more generally, we can ask about the possibility of solving m equations, depending upon parameters $x^1, \ldots ,x^n$, in m unknowns: If

$$f_i\colon \mathbf{R}^n \times \mathbf{R}^m \to \mathbf{R} \qquad i = 1, \ldots ,m$$

and

$$f_i(a^1, \ldots ,a^n, b^1, \ldots ,b^m) = 0 \qquad i = 1, \ldots ,m,$$

when can we find, for each $(x^1, \ldots ,x^n)$ near $(a^1, \ldots ,a^n)$ a unique $(y^1, \ldots ,y^m)$ near $(b^1, \ldots ,b^m)$ which satisfies $f_i(x^1, \ldots ,x^n,y^1, \ldots ,y^m) = 0$? The answer is provided by

2-12 Theorem (Implicit Function Theorem). *Suppose $f\colon \mathbf{R}^n \times \mathbf{R}^m \to \mathbf{R}^m$ is continuously differentiable in an open set containing (a,b) and $f(a,b) = 0$. Let M be the $m \times m$ matrix*

$$(D_{n+j}f^i(a,b)) \qquad 1 \le i, j \le m.$$

If $\det M \neq 0$, there is an open set $A \subset \mathbf{R}^n$ containing a and an open set $B \subset \mathbf{R}^m$ containing b, with the following property: for each $x \in A$ there is a unique $g(x) \in B$ such that $f(x,g(x)) = 0$. The function g is differentiable.

Proof. Define $F: \mathbf{R}^n \times \mathbf{R}^m \to \mathbf{R}^n \times \mathbf{R}^m$ by $F(x,y) = (x,f(x,y))$. Then $\det F'(a,b) = \det M \neq 0$. By Theorem 2-11 there is an open set $W \subset \mathbf{R}^n \times \mathbf{R}^m$ containing $F(a,b) = (a,0)$ and an open set in $\mathbf{R}^n \times \mathbf{R}^m$ containing (a,b), which we may take to be of the form $A \times B$, such that $F: A \times B \to W$ has a differentiable inverse $h: W \to A \times B$. Clearly h is of the form $h(x,y) = (x,k(x,y))$ for some differentiable function k (since F is of this form). Let $\pi: \mathbf{R}^n \times \mathbf{R}^m \to \mathbf{R}^m$ be defined by $\pi(x,y) = y$; then $\pi \circ F = f$. Therefore

$$\begin{aligned} f(x,k(x,y)) &= f \circ h(x,y) = (\pi \circ F) \circ h(x,y) \\ &= \pi \circ (F \circ h)(x,y) = \pi(x,y) = y. \end{aligned}$$

Thus $f(x,k(x,0)) = 0$; in other words we can define $g(x) = k(x,0)$. ▮

Since the function g is known to be differentiable, it is easy to find its derivative. In fact, since $f^i(x,g(x)) = 0$, taking D_j of both sides gives

$$0 = D_j f^i(x,g(x)) + \sum_{\alpha=1}^{m} D_{n+\alpha} f^i(x,g(x)) \cdot D_j g^\alpha(x) \qquad i,j = 1, \ldots, m.$$

Since $\det M \neq 0$, these equations can be solved for $D_j g^\alpha(x)$. The answer will depend on the various $D_j f^i(x,g(x))$, and therefore on $g(x)$. This is unavoidable, since the function g is not unique. Reconsidering the function $f: \mathbf{R}^2 \to \mathbf{R}$ defined by $f(x,y) = x^2 + y^2 - 1$, we note that two possible functions satisfying $f(x,g(x)) = 0$ are $g(x) = \sqrt{1 - x^2}$ and $g(x) = -\sqrt{1 - x^2}$. Differentiating $f(x,g(x)) = 0$ gives

$$D_1 f(x,g(x)) + D_2 f(x,g(x)) \cdot g'(x) = 0,$$

or

$$\begin{aligned} 2x + 2g(x) \cdot g'(x) &= 0, \\ g'(x) &= -x/g(x), \end{aligned}$$

which is indeed the case for either $g(x) = \sqrt{1 - x^2}$ or $g(x) = -\sqrt{1 - x^2}$.

A generalization of the argument for Theorem 2-12 can be given, which will be important in Chapter 5.

2-13 Theorem. *Let $f: \mathbf{R}^n \to \mathbf{R}^p$ be continuously differentiable in an open set containing a, where $p \leq n$. If $f(a) = 0$ and the $p \times n$ matrix $(D_jf^i(a))$ has rank p, then there is an open set $A \subset \mathbf{R}^n$ containing a and a differentiable function $h: A \to \mathbf{R}^n$ with differentiable inverse such that*

$$f \circ h(x^1, \ldots, x^n) = (x^{n-p+1}, \ldots, x^n).$$

Proof. We can consider f as a function $f: \mathbf{R}^{n-p} \times \mathbf{R}^p \to \mathbf{R}^p$. If det $M \neq 0$, then M is the $p \times p$ matrix $(D_{n-p+j}f^i(a))$, $1 \leq i, j \leq p$, then we are precisely in the situation considered in the proof of Theorem 2-12, and as we showed in that proof, there is h such that $f \circ h(x^1, \ldots, x^n) = (x^{n-p+1}, \ldots, x^n)$.

In general, since $(D_jf^i(a))$ has rank p, there will be $j_1 < \cdots < j_p$ such that the matrix $(D_jf^i(a))$ $1 \leq i \leq p$, $j = j_1, \ldots, j_p$ has non-zero determinant. If $g: \mathbf{R}^n \to \mathbf{R}^n$ permutes the x^j so that $g(x^1, \ldots, x^n) = (\ldots, x^{j_1}, \ldots, x^{j_p})$, then $f \circ g$ is a function of the type already considered, so $((f \circ g) \circ k)(x^1, \ldots, x^n) = (x^{n-p+1}, \ldots, x^n)$ for some k. Let $h = g \circ k$. ▮

Problems. 2-40. Use the implicit function theorem to re-do Problem 2-15(c).

2-41. Let $f: \mathbf{R} \times \mathbf{R} \to \mathbf{R}$ be differentiable. For each $x \in \mathbf{R}$ define $g_x: \mathbf{R} \to \mathbf{R}$ by $g_x(y) = f(x,y)$. Suppose that for each x there is a unique y with $g_x'(y) = 0$; let $c(x)$ be this y.

(a) If $D_{2,2}f(x,y) \neq 0$ for all (x,y), show that c is differentiable and

$$c'(x) = -\frac{D_{2,1}f(x,c(x))}{D_{2,2}f(x,c(x))}.$$

Hint: $g_x'(y) = 0$ can be written $D_2f(x,y) = 0$.

(b) Show that if $c'(x) = 0$, then for some y we have

$$D_{2,1}f(x,y) = 0,$$
$$D_2f(x,y) = 0.$$

(c) Let $f(x,y) = x(y \log y - y) - y \log x$. Find

$$\max_{\frac{1}{2} \leq x \leq 2} \left(\min_{\frac{1}{3} \leq y \leq 1} f(x,y) \right).$$

NOTATION

This section is a brief and not entirely unprejudiced discussion of classical notation connected with partial derivatives.

The partial derivative $D_1f(x,y,z)$ is denoted, among devotees of classical notation, by

$$\frac{\partial f(x,y,z)}{\partial x} \quad \text{or} \quad \frac{\partial f}{\partial x} \quad \text{or} \quad \frac{\partial f}{\partial x}(x,y,z) \quad \text{or} \quad \frac{\partial}{\partial x}f(x,y,z)$$

or any other convenient similar symbol. This notation forces one to write

$$\frac{\partial f}{\partial u}(u,v,w)$$

for $D_1f(u,v,w)$, although the symbol

$$\left.\frac{\partial f(x,y,z)}{\partial x}\right|_{(x,y,z)=(u,v,w)} \quad \text{or} \quad \frac{\partial f(x,y,z)}{\partial x}(u,v,w)$$

or something similar may be used (and must be used for an expression like $D_1f(7,3,2)$). Similar notation is used for D_2f and D_3f. Higher-order derivatives are denoted by symbols like

$$D_2D_1f(x,y,z) = \frac{\partial^2 f(x,y,z)}{\partial y\,\partial x}.$$

When $f\colon \mathbf{R} \to \mathbf{R}$, the symbol ∂ automatically reverts to d; thus

$$\frac{d\sin x}{dx}, \quad \text{not} \quad \frac{\partial \sin x}{\partial x}.$$

The mere statement of Theorem 2-2 in classical notation requires the introduction of irrelevant letters. The usual evaluation for $D_1(f \circ (g,h))$ runs as follows:

If $f(u,v)$ is a function and $u = g(x,y)$ and $v = h(x,y)$, then

$$\frac{\partial f(g(x,y),\, h(x,y))}{\partial x} = \frac{\partial f(u,v)}{\partial u}\frac{\partial u}{\partial x} + \frac{\partial f(u,v)}{\partial v}\frac{\partial v}{\partial x}.$$

[The symbol $\partial u/\partial x$ means $\partial/\partial x\, g(x,y)$ and $\partial/\partial u\, f(u,v)$ means

$D_1f(u,v) = D_1f(g(x,y), h(x,y))$.] This equation is often written simply

$$\frac{\partial f}{\partial x} = \frac{\partial f}{\partial u}\frac{\partial u}{\partial x} + \frac{\partial f}{\partial v}\frac{\partial v}{\partial x}.$$

Note that f means something different on the two sides of the equation!

The notation df/dx, always a little too tempting, has inspired many (usually meaningless) definitions of dx and df separately, the sole purpose of which is to make the equation

$$df = \frac{df}{dx} \cdot dx$$

work out. If $f\colon \mathbf{R}^2 \to \mathbf{R}$ then df is *defined*, classically, as

$$df = \frac{\partial f}{\partial x} dx + \frac{\partial f}{\partial y} dy$$

(whatever dx and dy mean).

Chapter 4 contains rigorous definitions which enable us to prove the above equations as theorems. It is a touchy question whether or not these modern definitions represent a real improvement over classical formalism; this the reader must decide for himself.

3

Integration

BASIC DEFINITIONS

The definition of the integral of a function $f: A \to \mathbf{R}$, where $A \subset \mathbf{R}^n$ is a closed rectangle, is so similar to that of the ordinary integral that a rapid treatment will be given.

Recall that a partition P of a closed interval $[a,b]$ is a sequence $t_0, \ldots, t_k$, where $a = t_0 \leq t_1 \leq \cdots \leq t_k = b$. The partition P divides the interval $[a,b]$ into k subintervals $[t_{i-1},t_i]$. A **partition** of a rectangle $[a_1,b_1] \times \cdots \times [a_n,b_n]$ is a collection $P = (P_1, \ldots, P_n)$, where each P_i is a partition of the interval $[a_i,b_i]$. Suppose, for example, that $P_1 = t_0, \ldots, t_k$ is a partition of $[a_1,b_1]$ and $P_2 = s_0, \ldots, s_l$ is a partition of $[a_2,b_2]$. Then the partition $P = (P_1,P_2)$ of $[a_1,b_1] \times [a_2,b_2]$ divides the closed rectangle $[a_1,b_1] \times [a_2,b_2]$ into $k \cdot l$ subrectangles, a typical one being $[t_{i-1},t_i] \times [s_{j-1},s_j]$. In general, if P_i divides $[a_i,b_i]$ into N_i subintervals, then $P = (P_1, \ldots, P_n)$ divides $[a_1,b_1] \times \cdots \times [a_n,b_n]$ into $N = N_1 \cdot \ldots \cdot N_n$ subrectangles. These subrectangles will be called **subrectangles of the partition** P.

Suppose now that A is a rectangle, $f: A \to \mathbf{R}$ is a bounded function, and P is a partition of A. For each subrectangle S of the partition let

$$m_S(f) = \inf\{f(x): x \in S\},$$
$$M_S(f) = \sup\{f(x): x \in S\},$$

and let $v(S)$ be the volume of S [the **volume** of a rectangle $[a_1,b_1] \times \cdots \times [a_n,b_n]$, and also of $(a_1,b_1) \times \cdots \times (a_n,b_n)$, is defined as $(b_1 - a_1) \cdot \ldots \cdot (b_n - a_n)$]. The **lower** and **upper sums** of f for P are defined by

$$L(f,P) = \sum_S m_S(f) \cdot v(S) \quad \text{and} \quad U(f,P) = \sum_S M_S(f) \cdot v(S).$$

Clearly $L(f,P) \leq U(f,P)$, and an even stronger assertion (3-2) is true.

3-1 Lemma. *Suppose the partition P' refines P (that is, each subrectangle of P' is contained in a subrectangle of P). Then*

$$L(f,P) \leq L(f,P') \quad \textit{and} \quad U(f,P') \leq U(f,P).$$

Proof. Each subrectangle S of P is divided into several subrectangles $S_1, \ldots, S_\alpha$ of P', so $v(S) = v(S_1) + \cdots + v(S_\alpha)$. Now $m_S(f) \leq m_{S_i}(f)$, since the values $f(x)$ for $x \in S$ include all values $f(x)$ for $x \in S_i$ (and possibly smaller ones). Thus

$$\begin{aligned} m_S(f) \cdot v(S) &= m_S(f) \cdot v(S_1) + \cdots + m_S(f) \cdot v(S_\alpha) \\ &\leq m_{S_1}(f) \cdot v(S_1) + \cdots + m_{S_\alpha}(f) \cdot v(S_\alpha). \end{aligned}$$

The sum, for all S, of the terms on the left side is $L(f,P)$, while the sum of all the terms on the right side is $L(f,P')$. Hence $L(f,P) \leq L(f,P')$. The proof for upper sums is similar. ▌

3-2 Corollary. *If P and P' are any two partitions, then $L(f,P') \leq U(f,P)$.*

Proof. Let P'' be a partition which refines both P and P'. (For example, let $P'' = (P_1'', \ldots, P_n'')$, where P_i'' is a partition of $[a_i,b_i]$ which refines both P_i and P'_i.) Then

$$L(f,P') \leq L(f,P'') \leq U(f,P'') \leq U(f,P). \quad \blacksquare$$

It follows from Corollary 3-2 that the least upper bound of all lower sums for f is less than or equal to the greatest lower bound of all upper sums for f. A function $f: A \to \mathbf{R}$ is called **integrable** on the rectangle A if f is bounded and $\sup\{L(f,P)\} = \inf\{U(f,P)\}$. This common number is then denoted $\int_A f$, and called the **integral** of f over A. Often, the notation $\int_A f(x^1, \ldots, x^n)dx^1 \cdots dx^n$ is used. If $f: [a,b] \to \mathbf{R}$, where $a \leq b$, then $\int_a^b f = \int_{[a,b]} f$. A simple but useful criterion for integrability is provided by

3-3 Theorem. *A bounded function $f: A \to \mathbf{R}$ is integrable if and only if for every $\varepsilon > 0$ there is a partition P of A such that $U(f,P) - L(f,P) < \varepsilon$.*

Proof. If this condition holds, it is clear that $\sup\{L(f,P)\} = \inf\{U(f,P)\}$ and f is integrable. On the other hand, if f is integrable, so that $\sup\{L(f,P)\} = \inf\{U(f,P)\}$, then for any $\varepsilon > 0$ there are partitions P and P' with $U(f,P) - L(f,P') < \varepsilon$. If P'' refines both P and P', it follows from Lemma 3-1 that $U(f,P'') - L(f,P'') \leq U(f,P) - L(f,P') < \varepsilon$. $\blacksquare$

In the following sections we will characterize the integrable functions and discover a method of computing integrals. For the present we consider two functions, one integrable and one not.

1. Let $f: A \to \mathbf{R}$ be a constant function, $f(x) = c$. Then for any partition P and subrectangle S we have $m_S(f) = M_S(f) = c$, so that $L(f,P) = U(f,P) = \Sigma_S c \cdot v(S) = c \cdot v(A)$. Hence $\int_A f = c \cdot v(A)$.

2. Let $f: [0,1] \times [0,1] \to \mathbf{R}$ be defined by

$$f(x,y) = \begin{cases} 0 & \text{if } x \text{ is rational,} \\ 1 & \text{if } x \text{ is irrational.} \end{cases}$$

If P is a partition, then every subrectangle S will contain points (x,y) with x rational, and also points (x,y) with x

irrational. Hence $m_S(f) = 0$ and $M_S(f) = 1$, so

$$L(f,P) = \sum_S 0 \cdot v(S) = 0$$

and

$$U(f,P) = \sum_S 1 \cdot v(S) = v([0,1] \times [0,1]) = 1.$$

Therefore f is not integrable.

Problems. **3-1.** Let $f\colon [0,1] \times [0,1] \to \mathbf{R}$ be defined by

$$f(x,y) = \begin{cases} 0 & \text{if } 0 \le x < \frac{1}{2}, \\ 1 & \text{if } \frac{1}{2} \le x \le 1. \end{cases}$$

Show that f is integrable and $\int_{[0,1]\times[0,1]} f = \frac{1}{2}$.

3-2. Let $f\colon A \to \mathbf{R}$ be integrable and let $g = f$ except at finitely many points. Show that g is integrable and $\int_A f = \int_A g$.

3-3. Let $f,g\colon A \to \mathbf{R}$ be integrable.

(a) For any partition P of A and subrectangle S, show that

$$m_S(f) + m_S(g) \le m_S(f+g) \quad \text{and} \quad M_S(f+g) \le M_S(f) + M_S(g)$$

and therefore

$$L(f,P) + L(g,P) \le L(f+g,\, P) \quad \text{and} \quad U(f+g,\, P) \le U(f,P) + U(g,P).$$

(b) Show that $f + g$ is integrable and $\int_A f + g = \int_A f + \int_A g$.

(c) For any constant c, show that $\int_A cf = c\int_A f$.

3-4. Let $f\colon A \to \mathbf{R}$ and let P be a partition of A. Show that f is integrable if and only if for each subrectangle S the function $f|S$, which consists of f restricted to S, is integrable, and that in this case $\int_A f = \Sigma_S \int_S f|S$.

3-5. Let $f,g\colon A \to \mathbf{R}$ be integrable and suppose $f \le g$. Show that $\int_A f \le \int_A g$.

3-6. If $f\colon A \to \mathbf{R}$ is integrable, show that $|f|$ is integrable and $\left|\int_A f\right| \le \int_A |f|$.

3-7. Let $f\colon [0,1] \times [0,1] \to \mathbf{R}$ be defined by

$$f(x,y) = \begin{cases} 0 & x \text{ irrational}, \\ 0 & x \text{ rational, } y \text{ irrational}, \\ 1/q & x \text{ rational, } y = p/q \text{ in lowest terms}. \end{cases}$$

Show that f is integrable and $\int_{[0,1]\times[0,1]} f = 0$.

MEASURE ZERO AND CONTENT ZERO

A subset A of $\mathbf{R}^n$ has (n-dimensional) **measure** 0 if for every $\varepsilon > 0$ there is a cover $\{U_1, U_2, U_3, \ldots\}$ of A by closed rectangles such that $\sum_{i=1}^{\infty} v(U_i) < \varepsilon$. It is obvious (but nevertheless useful to remember) that if A has measure 0 and $B \subset A$, then B has measure 0. The reader may verify that open rectangles may be used instead of closed rectangles in the definition of measure 0.

A set with only finitely many points clearly has measure 0. If A has infinitely many points which can be arranged in a sequence $a_1, a_2, a_3, \ldots$, then A also has measure 0, for if $\varepsilon > 0$, we can choose U_i to be a closed rectangle containing a_i with $v(U_i) < \varepsilon/2^i$. Then $\sum_{i=1}^{\infty} v(U_i) < \sum_{i=1}^{\infty} \varepsilon/2^i = \varepsilon$.

The set of all rational numbers between 0 and 1 is an important and rather surprising example of an infinite set whose members can be arranged in such a sequence. To see that this is so, list the fractions in the following array in the order indicated by the arrows (deleting repetitions and numbers greater than 1):

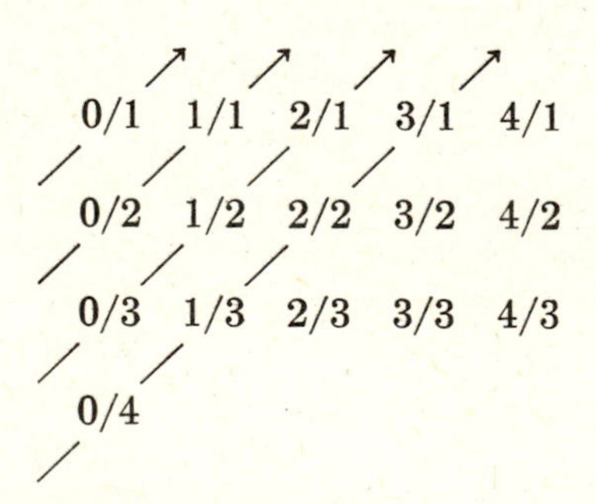

An important generalization of this idea can be given.

3-4 *Theorem.* *If* $A = A_1 \cup A_2 \cup A_3 \cup \cdots$ *and each* A_i *has measure* 0, *then* A *has measure* 0.

Proof. Let $\varepsilon > 0$. Since A_i has measure 0, there is a cover $\{U_{i,1}, U_{i,2}, U_{i,3}, \ldots\}$ of A_i by closed rectangles such that $\sum_{j=1}^{\infty} v(U_{i,j}) < \varepsilon/2^i$. Then the collection of all $U_{i,j}$ is a cover

of A. By considering the array

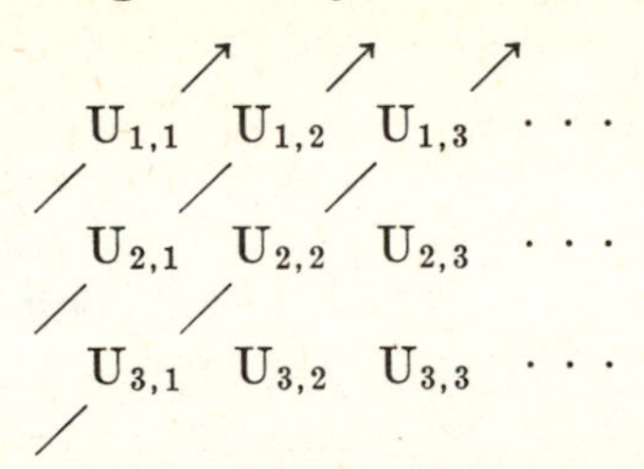

we see that this collection can be arranged in a sequence $V_1, V_2, V_3, \ldots$. Clearly $\Sigma_{i=1}^{\infty} v(V_i) < \Sigma_{i=1}^{\infty} \varepsilon/2^i = \varepsilon$. ∎

A subset A of $\mathbf{R}^n$ has (n-dimensional) **content** 0 if for every $\varepsilon > 0$ there is a *finite* cover $\{U_1, \ldots, U_n\}$ of A by closed rectangles such that $\Sigma_{i=1}^{n} v(U_i) < \varepsilon$. If A has content 0, then A clearly has measure 0. Again, open rectangles could be used instead of closed rectangles in the definition.

3-5 Theorem. *If $a < b$, then $[a,b] \subset \mathbf{R}$ does* not *have content* 0. *In fact, if $\{U_1, \ldots, U_n\}$ is a finite cover of $[a,b]$ by closed intervals, then $\Sigma_{i=1}^{n} v(U_i) \geq b - a$.*

Proof. Clearly we can assume that each $U_i \subset [a,b]$. Let $a = t_0 < t_1 < \ldots < t_k = b$ be all endpoints of all U_i. Then each $v(U_i)$ is the sum of certain $t_j - t_{j-1}$. Moreover, each $[t_{j-1},t_j]$ lies in at least one U_i (namely, any one which contains an interior point of $[t_{j-1},t_j]$), so $\Sigma_{i=1}^{n} v(U_i) \geq \Sigma_{j=1}^{k} (t_j - t_{j-1}) = b - a$. ∎

If $a < b$, it is also true that $[a,b]$ does not have measure 0. This follows from

3-6 Theorem. *If A is compact and has measure* 0, *then A has content* 0.

Proof. Let $\varepsilon > 0$. Since A has measure 0, there is a cover $\{U_1, U_2, \ldots\}$ of A by open rectangles such that $\Sigma_{i=1}^{\infty} v(U_i)$

$< \varepsilon$. Since A is compact, a finite number $U_1, \ldots, U_n$ of the U_i also cover A and surely $\sum_{i=1}^{n} v(U_i) < \varepsilon$. ∎

The conclusion of Theorem 3-6 is false if A is not compact. For example, let A be the set of rational numbers between 0 and 1; then A has measure 0. Suppose, however, that $\{[a_1,b_1], \ldots, [a_n,b_n]\}$ covers A. Then A is contained in the closed set $[a_1,b_1] \cup \cdots \cup [a_n,b_n]$, and therefore $[0,1] \subset [a_1,b_1] \cup \cdots \cup [a_n,b_n]$. It follows from Theorem 3-5 that $\sum_{i=1}^{n}(b_i - a_i) \geq 1$ for any such cover, and consequently A does not have content 0.

Problems. **3-8.** Prove that $[a_1,b_1] \times \cdots \times [a_n,b_n]$ does not have content 0 if $a_i < b_i$ for each i.

3-9. (a) Show that an unbounded set cannot have content 0.
(b) Give an example of a closed set of measure 0 which does not have content 0.

3-10. (a) If C is a set of content 0, show that the boundary of C has content 0.
(b) Give an example of a bounded set C of measure 0 such that the boundary of C does not have measure 0.

3-11. Let A be the set of Problem 1-18. If $\sum_{i=1}^{\infty}(b_i - a_i) < 1$, show that the boundary of A does not have measure 0.

3-12. Let $f\colon [a,b] \to \mathbf{R}$ be an increasing function. Show that $\{x\colon f \text{ is discontinuous at } x\}$ has measure 0. *Hint:* Use Problem 1-30 to show that $\{x\colon o(f,x) > 1/n\}$ is finite, for each integer n.

3-13.* (a) Show that the collection of all rectangles $[a_1,b_1] \times \cdots \times [a_n,b_n]$ with all a_i and b_i rational can be arranged in a sequence.
(b) If $A \subset \mathbf{R}^n$ is any set and $\mathcal{O}$ is an open cover of A, show that there is a sequence $U_1, U_2, U_3, \ldots$ of members of $\mathcal{O}$ which also cover A. *Hint:* For each $x \in A$ there is a rectangle $B = [a_1,b_1] \times \cdots \times [a_n,b_n]$ with all a_i and b_i rational such that $x \in B \subset U$ for some $U \in \mathcal{O}$.

INTEGRABLE FUNCTIONS

Recall that $o(f,x)$ denotes the oscillation of f at x.

3-7 Lemma. *Let A be a closed rectangle and let $f\colon A \to \mathbf{R}$ be a bounded function such that $o(f,x) < \varepsilon$ for all $x \in A$. Then there is a partition P of A with $U(f,P) - L(f,P) < \varepsilon \cdot v(A)$.*

Proof. For each $x \in A$ there is a closed rectangle U_x, containing x in its interior, such that $M_{U_x}(f) - m_{U_x}(f) < \varepsilon$. Since A is compact, a finite number $U_{x_1}, \ldots, U_{x_n}$ of the sets U_x cover A. Let P be a partition for A such that each subrectangle S of P is contained in some U_{x_i}. Then $M_S(f) - m_S(f) < \varepsilon$ for each subrectangle S of P, so that $U(f,P) - L(f,P) = \Sigma_S[M_S(f) - m_S(f)] \cdot v(S) < \varepsilon \cdot v(A)$. ▮

3-8 Theorem. *Let A be a closed rectangle and $f: A \to \mathbf{R}$ a bounded function. Let $B = \{x: f$ is not continuous at $x\}$. Then f is integrable if and only if B is a set of measure* 0.

Proof. Suppose first that B has measure 0. Let $\varepsilon > 0$ and let $B_\varepsilon = \{x: o(f,x) \geq \varepsilon\}$. Then $B_\varepsilon \subset B$, so that B_ε has measure 0. Since (Theorem 1-11) B_ε is compact, B_ε has content 0. Thus there is a finite collection $U_1, \ldots, U_n$ of closed rectangles, whose interiors cover B_ε, such that $\Sigma_{i=1}^n v(U_i) < \varepsilon$. Let P be a partition of A such that every subrectangle S of P is in one of two groups (see Figure 3-1):

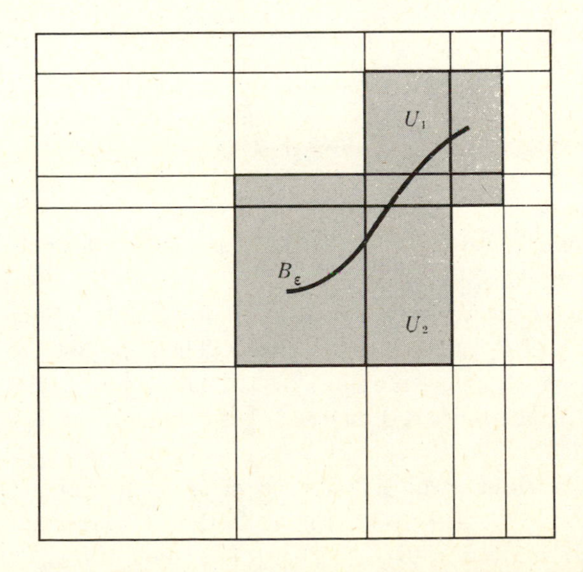

FIGURE 3-1. *The shaded rectangles are in* $\mathcal{S}_1$.

(1) $\mathcal{S}_1$, which consists of subrectangles S, such that $S \subset U_i$ for some i.

(2) $\mathcal{S}_2$, which consists of subrectangles S with $S \cap B_\varepsilon = \emptyset$.

Let $|f(x)| < M$ for $x \in A$. Then $M_S(f) - m_S(f) < 2M$ for every S. Therefore

$$\sum_{S \in \mathcal{S}_1} [M_S(f) - m_S(f)] \cdot v(S) < 2M \sum_{i=1}^{n} v(U_i) < 2M\varepsilon.$$

Now, if $S \in \mathcal{S}_2$, then $o(f,x) < \varepsilon$ for $x \in S$. Lemma 3-7 implies that there is a refinement P' of P such that

$$\sum_{S' \subset S} [M_{S'}(f) - m_{S'}(f)] \cdot v(S') < \varepsilon \cdot v(S)$$

for $S \in \mathcal{S}_2$. Then

$$\begin{aligned} U(f,P') - L(f,P') &= \sum_{S' \subset S \in \mathcal{S}_1} [M_{S'}(f) - m_{S'}(f)] \cdot v(S') \\ &\quad + \sum_{S' \subset S \in \mathcal{S}_2} [M_{S'}(f) - m_{S'}(f)] \cdot v(S') \\ &< 2M\varepsilon + \sum_{S \in \mathcal{S}_2} \varepsilon \cdot v(S) \\ &\leq 2M\varepsilon + \varepsilon \cdot v(A). \end{aligned}$$

Since M and $v(A)$ are fixed, this shows that we can find a partition P' with $U(f,P') - L(f,P')$ as small as desired. Thus f is integrable.

Suppose, conversely, that f is integrable. Since $B = B_1 \cup B_{\frac{1}{2}} \cup B_{\frac{1}{3}} \cup \cdots$, it suffices (Theorem 3-4) to prove that each $B_{1/n}$ has measure 0. In fact we will show that each $B_{1/n}$ has content 0 (since $B_{1/n}$ is compact, this is actually equivalent).

If $\varepsilon > 0$, let P be a partition of A such that $U(f,P) - L(f,P) < \varepsilon/n$. Let $\mathcal{S}$ be the collection of subrectangles S of P which intersect $B_{1/n}$. Then $\mathcal{S}$ is a cover of $B_{1/n}$. Now if

$S \in \mathcal{S}$, then $M_S(f) - m_S(f) \geq 1/n$. Thus

$$\begin{aligned}\frac{1}{n} \cdot \sum_{S \in \mathcal{S}} v(S) &\leq \sum_{S \in \mathcal{S}} [M_S(f) - m_S(f)] \cdot v(S) \\ &\leq \sum_{S} [M_S(f) - m_S(f)] \cdot v(S) \\ &< \frac{\varepsilon}{n},\end{aligned}$$

and consequently $\Sigma_{S \in \mathcal{S}} v(S) < \varepsilon$. ▌

We have thus far dealt only with the integrals of functions over rectangles. Integrals over other sets are easily reduced to this type. If $C \subset \mathbf{R}^n$, the **characteristic function** χ_C of C is defined by

$$\chi_C(x) = \begin{cases} 0 & x \notin C, \\ 1 & x \in C. \end{cases}$$

If $C \subset A$ for some closed rectangle A and $f\colon A \to \mathbf{R}$ is bounded, then $\int_C f$ is defined as $\int_A f \cdot \chi_C$, provided $f \cdot \chi_C$ is integrable. This certainly occurs (Problem 3-14) if f and χ_C are integrable.

3-9 Theorem. *The function $\chi_C\colon A \to \mathbf{R}$ is integrable if and only if the boundary of C has measure* 0 (*and hence content* 0).

Proof. If x is in the interior of C, then there is an open rectangle U with $x \in U \subset C$. Thus $\chi_C = 1$ on U and χ_C is clearly continuous at x. Similarly, if x is in the exterior of C, there is an open rectangle U with $x \in U \subset \mathbf{R}^n - C$. Hence $\chi_C = 0$ on U and χ_C is continuous at x. Finally, if x is in the boundary of C, then for every open rectangle U containing x, there is $y_1 \in U \cap C$, so that $\chi_C(y_1) = 1$ and there is $y_2 \in U \cap (\mathbf{R}^n - C)$, so that $\chi_C(y_2) = 0$. Hence χ_C is not continuous at x. Thus $\{x\colon \chi_C \text{ is not continuous at } x\} =$ boundary C, and the result follows from Theorem 3-8. ▌

A bounded set C whose boundary has measure 0 is called **Jordan-measurable.** The integral $\int_C 1$ is called the (n-dimensional) **content** of C, or the (n-dimensional) **volume** of C. Naturally one-dimensional volume is often called **length,** and two-dimensional volume, **area.**

Problem 3-11 shows that even an open set C may not be Jordan-measurable, so that $\int_C f$ is not necessarily defined even if C is open and f is continuous. This unhappy state of affairs will be rectified soon.

Problems. **3-14.** Show that if $f,g\colon A \to \mathbf{R}$ are integrable, so is $f \cdot g$.

3-15. Show that if C has content 0, then $C \subset A$ for some closed rectangle A and C is Jordan-measurable and $\int_A \chi_C = 0$.

3-16. Give an example of a bounded set C of measure 0 such that $\int_A \chi_C$ does not exist.

3-17. If C is a bounded set of measure 0 and $\int_A \chi_C$ exists, show that $\int_A \chi_C = 0$. *Hint:* Show that $L(f,P) = 0$ for all partitions P. Use Problem 3-8.

3-18. If $f\colon A \to \mathbf{R}$ is non-negative and $\int_A f = 0$, show that $\{x\colon f(x) \neq 0\}$ has measure 0. *Hint:* Prove that $\{x\colon f(x) > 1/n\}$ has content 0.

3-19. Let U be the open set of Problem 3-11. Show that if $f = \chi_U$ except on a set of measure 0, then f is not integrable on $[0,1]$.

3-20. Show that an increasing function $f\colon [a,b] \to \mathbf{R}$ is integrable on $[a,b]$.

3-21. If A is a closed rectangle, show that $C \subset A$ is Jordan-measurable if and only if for every $\varepsilon > 0$ there is a partition P of A such that $\Sigma_{S \in \mathcal{S}_1} v(S) - \Sigma_{S \in \mathcal{S}_2} v(S) < \varepsilon$, where $\mathcal{S}_1$ consists of all subrectangles intersecting C and $\mathcal{S}_2$ all subrectangles contained in C.

3-22.* If A is a Jordan-measurable set and $\varepsilon > 0$, show that there is a compact Jordan-measurable set $C \subset A$ such that $\int_{A-C} 1 < \varepsilon$.

FUBINI'S THEOREM

The problem of calculating integrals is solved, in some sense, by Theorem 3-10, which reduces the computation of integrals over a closed rectangle in $\mathbf{R}^n$, $n > 1$, to the computation of integrals over closed intervals in $\mathbf{R}$. Of sufficient importance to deserve a special designation, this theorem is usually referred to as Fubini's theorem, although it is more or less a special case of a theorem proved by Fubini long after Theorem 3-10 was known.

The idea behind the theorem is best illustrated (Figure 3-2) for a positive continuous function $f\colon [a,b] \times [c,d] \to \mathbf{R}$. Let $t_0, \ldots, t_n$ be a partition of $[a,b]$ and divide $[a,b] \times [c,d]$ into n strips by means of the line segments $\{t_i\} \times [c,d]$. If g_x is defined by $g_x(y) = f(x,y)$, then the area of the region under the graph of f and above $\{x\} \times [c,d]$ is

$$\int_c^d g_x = \int_c^d f(x,y)dy.$$

The volume of the region under the graph of f and above $[t_{i-1},t_i] \times [c,d]$ is therefore approximately equal to $(t_i - t_{i-1}) \cdot \int_c^d f(x,y)dy$, for any $x \in [t_{i-1},t_i]$. Thus

$$\int_{[a,b]\times[c,d]} f = \sum_{i=1}^{n} \int_{[t_{i-1},t_i]\times[c,d]} f$$

is approximately $\sum_{i=1}^{n}(t_i - t_{i-1}) \cdot \int_c^d f(x_i,y)dy$, with x_i in

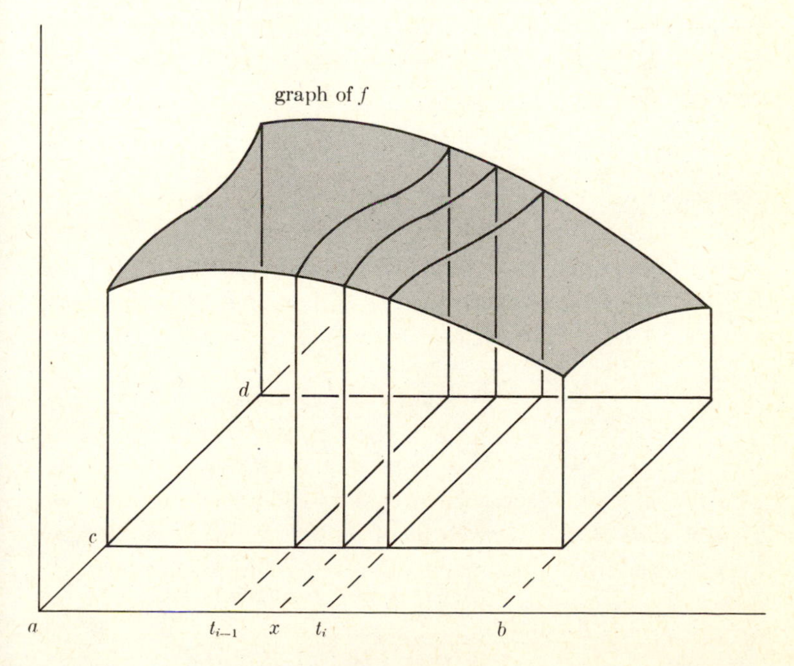

FIGURE 3-2

$[t_{i-1},t_i]$. On the other hand, sums similar to these appear in the definition of $\int_a^b(\int_c^d f(x,y)dy)dx$. Thus, if h is defined by $h(x) = \int_c^d g_x = \int_c^d f(x,y)dy$, it is reasonable to hope that h is integrable on $[a,b]$ and that

$$\int_{[a,b]\times[c,d]} f = \int_a^b h = \int_a^b \left(\int_c^d f(x,y)dy\right) dx.$$

This will indeed turn out to be true when f is continuous, but in the general case difficulties arise. Suppose, for example, that the set of discontinuities of f is $\{x_0\} \times [c,d]$ for some $x_0 \in [a,b]$. Then f is integrable on $[a,b] \times [c,d]$ but $h(x_0) = \int_c^d f(x_0,y)dy$ may not even be defined. The statement of Fubini's theorem therefore looks a little strange, and will be followed by remarks about various special cases where simpler statements are possible.

We will need one bit of terminology. If $f\colon A \to \mathbf{R}$ is a bounded function on a closed rectangle, then, whether or not f is integrable, the least upper bound of all lower sums, and the greatest lower bound of all upper sums, both exist. They are called the **lower** and **upper integrals** of f on A, and denoted

$$\mathbf{L}\int_A f \qquad \text{and} \qquad \mathbf{U}\int_A f,$$

respectively.

3-10 Theorem (Fubini's Theorem). *Let $A \subset \mathbf{R}^n$ and $B \subset \mathbf{R}^m$ be closed rectangles, and let $f\colon A \times B \to \mathbf{R}$ be integrable. For $x \in A$ let $g_x\colon B \to \mathbf{R}$ be defined by $g_x(y) = f(x,y)$ and let*

$$\mathfrak{L}(x) = \mathbf{L}\int_B g_x = \mathbf{L}\int_B f(x,y)dy,$$
$$\mathfrak{U}(x) = \mathbf{U}\int_B g_x = \mathbf{U}\int_B f(x,y)dy.$$

Then $\mathfrak{L}$ and $\mathfrak{U}$ are integrable on A and

$$\int_{A\times B} f = \int_A \mathfrak{L} = \int_A \left(\mathbf{L}\int_B f(x,y)dy\right) dx,$$
$$\int_{A\times B} f = \int_A \mathfrak{U} = \int_A \left(\mathbf{U}\int_B f(x,y)dy\right) dx.$$

(The integrals on the right side are called **iterated integrals** for f.)

Proof. Let P_A be a partition of A and P_B a partition of B. Together they give a partition P of $A \times B$ for which any subrectangle S is of the form $S_A \times S_B$, where S_A is a subrectangle of the partition P_A, and S_B is a subrectangle of the partition P_B. Thus

$$\begin{aligned} L(f,P) = \sum_S m_S(f) \cdot v(S) &= \sum_{S_A,S_B} m_{S_A \times S_B}(f) \cdot v(S_A \times S_B) \\ &= \sum_{S_A} \Big(\sum_{S_B} m_{S_A \times S_B}(f) \cdot v(S_B) \Big) \cdot v(S_A). \end{aligned}$$

Now, if $x \in S_A$, then clearly $m_{S_A \times S_B}(f) \leq m_{S_B}(g_x)$. Consequently, for $x \in S_A$ we have

$$\sum_{S_B} m_{S_A \times S_B}(f) \cdot v(S_B) \leq \sum_{S_B} m_{S_B}(g_x) \cdot v(S_B) \leq \mathbf{L} \int_B g_x = \mathfrak{L}(x).$$

Therefore

$$\sum_{S_A} \Big(\sum_{S_B} m_{S_A \times S_B}(f) \cdot v(S_B) \Big) \cdot v(S_A) \leq L(\mathfrak{L},P_A).$$

We thus obtain

$$L(f,P) \leq L(\mathfrak{L},P_A) \leq U(\mathfrak{L},P_A) \leq U(\mathfrak{U},P_A) \leq U(f,P),$$

where the proof of the last inequality is entirely analogous to the proof of the first. Since f is integrable, $\sup\{L(f,P)\} = \inf\{U(f,P)\} = \int_{A \times B} f$. Hence

$$\sup\{L(\mathfrak{L},P_A)\} = \inf\{U(\mathfrak{L},P_A)\} = \textstyle\int_{A \times B} f.$$

In other words, $\mathfrak{L}$ is integrable on A and $\int_{A \times B} f = \int_A \mathfrak{L}$. The assertion for $\mathfrak{U}$ follows similarly from the inequalities

$$L(f,P) \leq L(\mathfrak{L},P_A) \leq L(\mathfrak{U},P_A) \leq U(\mathfrak{U},P_A) \leq U(f,P). \quad \blacksquare$$

Remarks. 1. A similar proof shows that

$$\int_{A \times B} f = \int_B \Big(\mathbf{L} \int_A f(x,y)dx \Big)\, dy = \int_B \Big(\mathbf{U} \int_A f(x,y)dx \Big)\, dy.$$

These integrals are called *iterated integrals* for f in the reverse order from those of the theorem. As several problems show, the possibility of interchanging the orders of iterated integrals has many consequences.

2. In practice it is often the case that each g_x is integrable, so that $\int_{A\times B} f = \int_A(\int_B f(x,y)dy)dx$. This certainly occurs if f is continuous.

3. The worst irregularity commonly encountered is that g_x is not integrable for a finite number of $x \in A$. In this case $\mathfrak{L}(x) = \int_B f(x,y)dy$ for all but these finitely many x. Since $\int_A \mathfrak{L}$ remains unchanged if $\mathfrak{L}$ is redefined at a finite number of points, we can still write $\int_{A\times B} f = \int_A(\int_B f(x,y)dy)dx$, provided that $\int_B f(x,y)dy$ is defined arbitrarily, say as 0, when it does not exist.

4. There are cases when this will not work and Theorem 3-10 must be used as stated. Let $f\colon [0,1] \times [0,1] \to \mathbf{R}$ be defined by

$$f(x,y) = \begin{cases} 1 & \text{if } x \text{ is irrational,} \\ 1 & \text{if } x \text{ is rational and } y \text{ is irrational,} \\ 1 - 1/q & \text{if } x = p/q \text{ in lowest terms and } y \text{ is rational.} \end{cases}$$

Then f is integrable and $\int_{[0,1]\times[0,1]} f = 1$. Now $\int_0^1 f(x,y)dy = 1$ if x is irrational, and does not exist if x is rational. Therefore h is not integrable if $h(x) = \int_0^1 f(x,y)dy$ is set equal to 0 when the integral does not exist.

5. If $A = [a_1,b_1] \times \cdots \times [a_n,b_n]$ and $f\colon A \to \mathbf{R}$ is sufficiently nice, we can apply Fubini's theorem repeatedly to obtain

$$\int_A f = \int_{a_n}^{b_n}\left(\cdots\left(\int_{a_1}^{b_1} f(x^1, \ldots, x^n)dx^1\right)\cdots\right)dx^n.$$

6. If $C \subset A \times B$, Fubini's theorem can be used to evaluate $\int_C f$, since this is by definition $\int_{A\times B} \chi_C f$. Suppose, for example, that

$$C = [-1,1] \times [-1,1] - \{(x,y)\colon |(x,y)| < 1\}.$$

Then

$$\int_C f = \int_{-1}^{1}\left(\int_{-1}^{1} f(x,y)\cdot\chi_C(x,y)dy\right)dx.$$

Now

$$\chi_C(x,y) = \begin{cases} 1 & \text{if } y > \sqrt{1-x^2} \text{ or } y < -\sqrt{1-x^2}, \\ 0 & \text{otherwise.} \end{cases}$$

Therefore

$$\int_{-1}^{1} f(x,y) \cdot \chi_C(x,y)dy = \int_{-1}^{-\sqrt{1-x^2}} f(x,y)dy + \int_{\sqrt{1-x^2}}^{1} f(x,y)dy.$$

In general, if $C \subset A \times B$, the main difficulty in deriving expressions for $\int_C f$ will be determining $C \cap (\{x\} \times B)$ for $x \in A$. If $C \cap (A \times \{y\})$ for $y \in B$ is easier to determine, one should use the iterated integral

$$\int_C f = \int_B \left(\int_A f(x,y) \cdot \chi_C(x,y)dx \right) dy.$$

Problems. **3-23.** Let $C \subset A \times B$ be a set of content 0. Let $A' \subset A$ be the set of all $x \in A$ such that $\{y \in B\colon (x,y) \in C\}$ is *not* of content 0. Show that A' is a set of measure 0. *Hint:* χ_C is integrable and $\int_{A \times B} \chi_C = \int_A \mathcal{U} = \int_A \mathcal{L}$, so $\int_A \mathcal{U} - \mathcal{L} = 0$.

3-24. Let $C \subset [0,1] \times [0,1]$ be the union of all $\{p/q\} \times [0, 1/q]$, where p/q is a rational number in $[0,1]$ written in lowest terms. Use C to show that the word "measure" in Problem 3-23 cannot be replaced by "content."

3-25. Use induction on n to show that $[a_1,b_1] \times \cdots \times [a_n,b_n]$ is not a set of measure 0 (or content 0) if $a_i < b_i$ for each i.

3-26. Let $f\colon [a,b] \to \mathbf{R}$ be integrable and non-negative and let $A_f = \{(x,y)\colon a \le x \le b \text{ and } 0 \le y \le f(x)\}$. Show that A_f is Jordan-measurable and has area $\int_a^b f$.

3-27. If $f\colon [a,b] \times [a,b] \to \mathbf{R}$ is continuous, show that

$$\int_a^b \int_a^y f(x,y)dx\, dy = \int_a \int_x f(x,y)dy\, dx.$$

Hint: Compute $\int_C f$ in two different ways for a suitable set $C \subset [a,b] \times [a,b]$.

3-28.* Use Fubini's theorem to give an easy proof that $D_{1,2}f = D_{2,1}f$ if these are continuous. *Hint:* If $D_{1,2}f(a) - D_{2,1}f(a) > 0$, there is a rectangle A containing a such that $D_{1,2}f - D_{2,1}f > 0$ on A.

3-29. Use Fubini's theorem to derive an expression for the volume of a set of $\mathbf{R}^3$ obtained by revolving a Jordan-measurable set in the yz-plane about the z-axis.

3-30. Let C be the set in Problem 1-17. Show that

$$\int_{[0,1]}\left(\int_{[0,1]} \chi_C(x,y)dx\right)dy = \int_{[0,1]}\left(\int_{[0,1]} \chi_C(y,x)dy\right)dx = 0$$

but that $\int_{[0,1]\times[0,1]} \chi_C$ does not exist.

3-31. If $A = [a_1,b_1] \times \cdots \times [a_n,b_n]$ and $f\colon A \to \mathbf{R}$ is continuous, define $F\colon A \to \mathbf{R}$ by

$$F(x) = \int_{[a_1,x^1]\times\cdots\times[a_n,x^n]} f.$$

What is $D_iF(x)$, for x in the interior of A?

3-32.* Let $f\colon [a,b] \times [c,d] \to \mathbf{R}$ be continuous and suppose D_2f is continuous. Define $F(y) = \int_a^b f(x,y)dx$. Prove *Leibnitz's rule:* $F'(y) = \int_a^b D_2f(x,y)dx$. *Hint:* $F(y) = \int_a^b f(x,y)dx = \int_a^b(\int_c^y D_2f(x,y)dy + f(x,c))dx$. (The proof will show that continuity of D_2f may be replaced by considerably weaker hypotheses.)

3-33. If $f\colon [a,b] \times [c,d] \to \mathbf{R}$ is continuous and D_2f is continuous, define $F(x,y) = \int_a^x f(t,y)dt$.

(a) Find D_1F and D_2F.

(b) If $G(x) = \int_a^{g(x)} f(t,x)dt$, find $G'(x)$.

3-34.* Let $g_1,g_2\colon \mathbf{R}^2 \to \mathbf{R}$ be continuously differentiable and suppose $D_1g_2 = D_2g_1$. As in Problem 2-21, let

$$f(x,y) = \int_0^x g_1(t,0)dt + \int_0^y g_2(x,t)dt.$$

Show that $D_1f(x,y) = g_1(x,y)$.

3-35.* (*a*) Let $g\colon \mathbf{R}^n \to \mathbf{R}^n$ be a linear transformation of one of the following types:

$$\begin{cases} g(e_i) = e_i & i \neq j \\ g(e_j) = ae_j \end{cases}$$

$$\begin{cases} g(e_i) = e_i & i \neq j \\ g(e_j) = e_j + e_k \end{cases}$$

$$\begin{cases} g(e_k) = e_k & k \neq i, j \\ g(e_i) = e_j \\ g(e_j) = e_i. \end{cases}$$

If U is a rectangle, show that the volume of $g(U)$ is $|\det g| \cdot v(U)$.

(b) Prove that $|\det g| \cdot v(U)$ is the volume of $g(U)$ for any linear transformation $g\colon \mathbf{R}^n \to \mathbf{R}^n$. *Hint:* If $\det g \neq 0$, then g is the composition of linear transformations of the type considered in (a).

3-36. (Cavalieri's principle). Let A and B be Jordan-measurable subsets of $\mathbf{R}^3$. Let $A_c = \{(x,y)\colon (x,y,c) \in A\}$ and define B_c similarly. Suppose each A_c and B_c are Jordan-measurable and have the same area. Show that A and B have the same volume.

PARTITIONS OF UNITY

In this section we introduce a tool of extreme importance in the theory of integration.

3-11 Theorem. *Let $A \subset \mathbf{R}^n$ and let $\mathcal{O}$ be an open cover of A. Then there is a collection Φ of C^∞ functions φ defined in an open set containing A, with the following properties:*

(1) *For each $x \in A$ we have $0 \leq \varphi(x) \leq 1$.*
(2) *For each $x \in A$ there is an open set V containing x such that all but finitely many $\varphi \in \Phi$ are 0 on V.*
(3) *For each $x \in A$ we have $\Sigma_{\varphi \in \Phi} \varphi(x) = 1$* (by (2) for each x this sum is finite in some open set containing x).
(4) *For each $\varphi \in \Phi$ there is an open set U in $\mathcal{O}$ such that $\varphi = 0$ outside of some closed set contained in U.*

(A collection Φ satisfying (1) to (3) is called a C^∞ **partition of unity** for A. If Φ also satisfies (4), it is said to be **subordinate** to the cover $\mathcal{O}$. In this chapter we will only use continuity of the functions φ.)

Proof. *Case 1. A is compact.*

Then a finite number $U_1, \ldots, U_n$ of open sets in $\mathcal{O}$ cover A. It clearly suffices to construct a partition of unity subordinate to the cover $\{U_1, \ldots, U_n\}$. We will first find compact sets $D_i \subset U_i$ whose interiors cover A. The sets D_i are constructed inductively as follows. Suppose that $D_1, \ldots, D_k$ have been chosen so that {interior $D_1, \ldots,$ interior D_k, $U_{k+1}, \ldots, U_n$} covers A. Let

$$C_{k+1} = A - (\text{int } D_1 \cup \cdots \cup \text{int } D_k \cup U_{k+2} \cup \cdots \cup U_n).$$

Then $C_{k+1} \subset U_{k+1}$ is compact. Hence (Problem 1-22) we can find a compact set D_{k+1} such that

$$C_{k+1} \subset \text{interior } D_{k+1} \quad \text{and} \quad D_{k+1} \subset U_{k+1}.$$

Having constructed the sets $D_1, \ldots, D_n$, let ψ_i be a non-negative C^∞ function which is positive on D_i and 0 outside of some closed set contained in U_i (Problem 2-26). Since

$\{D_1, \ldots, D_n\}$ covers A, we have $\psi_1(x) + \cdots + \psi_n(x) > 0$ for all x in some open set U containing A. On U we can define

$$\varphi_i(x) = \frac{\psi_i(x)}{\psi_1(x) + \cdots + \psi_n(x)}.$$

If $f\colon U \to [0,1]$ is a C^∞ function which is 1 on A and 0 outside of some closed set in U, then $\Phi = \{f \cdot \varphi_1, \ldots, f \cdot \varphi_n\}$ is the desired partition of unity.

Case 2. $A = A_1 \cup A_2 \cup A_3 \cup \cdots$, *where each* A_i *is compact and* $A_i \subset$ *interior* A_{i+1}.

For each i let $\mathcal{O}_i$ consist of all $U \cap$ (interior $A_{i+1} - A_{i-2}$) for U in $\mathcal{O}$. Then $\mathcal{O}_i$ is an open cover of the compact set $B_i = A_i -$ interior A_{i-1}. By case 1 there is a partition of unity Φ_i for B_i, subordinate to $\mathcal{O}_i$. For each $x \in A$ the sum

$$\sigma(x) = \sum_{\varphi \in \Phi_i,\ \text{all}\ i} \varphi(x)$$

is a finite sum in some open set containing x, since if $x \in A_i$ we have $\varphi(x) = 0$ for $\varphi \in \Phi_j$ with $j \geq i + 2$. For each φ in each Φ_i, define $\varphi'(x) = \varphi(x)/\sigma(x)$. The collection of all φ' is the desired partition of unity.

Case 3. A *is open.*

Let $A_i =$

$$\{x \in A\colon |x| \leq i \text{ and distance from } x \text{ to boundary } A \geq 1/i\},$$

and apply case 2.

Case 4. A *is arbitrary.*

Let B be the union of all U in $\mathcal{O}$. By case 3 there is a partition of unity for B; this is also a partition of unity for A. ▌

An important consequence of condition (2) of the theorem should be noted. Let $C \subset A$ be compact. For each $x \in C$ there is an open set V_x containing x such that only finitely many $\varphi \in \Phi$ are not 0 on V_x. Since C is compact, finitely many such V_x cover C. Thus only finitely many $\varphi \in \Phi$ are not 0 on C.

One important application of partitions of unity will illustrate their main role—piecing together results obtained locally.

An open cover $\mathcal{O}$ of an open set $A \subset \mathbf{R}^n$ is **admissible** if each $U \in \mathcal{O}$ is contained in A. If Φ is subordinate to $\mathcal{O}$, $f\colon A \to \mathbf{R}$ is bounded in some open set around each point of A, and $\{x\colon f \text{ is discontinuous at } x\}$ has measure 0, then each $\int_A \varphi \cdot |f|$ exists. We define f to be **integrable** (in the extended sense) if $\Sigma_{\varphi\in\Phi}\int_A \varphi \cdot |f|$ converges (the proof of Theorem 3-11 shows that the φ's may be arranged in a sequence). This implies convergence of $\Sigma_{\varphi\in\Phi}|\int_A \varphi \cdot f|$, and hence absolute convergence of $\Sigma_{\varphi\in\Phi}\int_A \varphi \cdot f$, which we define to be $\int_A f$. These definitions do not depend on $\mathcal{O}$ or Φ (but see Problem 3-38).

3-12 Theorem.

(1) *If Ψ is another partition of unity, subordinate to an admissible cover $\mathcal{O}'$ of A, then $\Sigma_{\psi\in\Psi}\int_A \psi \cdot |f|$ also converges, and*

$$\sum_{\varphi\in\Phi}\int_A \varphi \cdot f = \sum_{\psi\in\Psi}\int_A \psi \cdot f.$$

(2) *If A and f are bounded, then f is integrable in the extended sense.*

(3) *If A is Jordan-measurable and f is bounded, then this definition of $\int_A f$ agrees with the old one.*

Proof

(1) Since $\varphi \cdot f = 0$ except on some compact set C, and there are only finitely many ψ which are non-zero on C, we can write

$$\sum_{\varphi\in\Phi}\int_A \varphi \cdot f = \sum_{\varphi\in\Phi}\int_A \sum_{\psi\in\Psi} \psi \cdot \varphi \cdot f = \sum_{\varphi\in\Phi}\sum_{\psi\in\Psi}\int_A \psi \cdot \varphi \cdot f.$$

This result, applied to $|f|$, shows the convergence of $\Sigma_{\varphi\in\Phi}\Sigma_{\psi\in\Psi}\int_A \psi \cdot \varphi \cdot |f|$, and hence of $\Sigma_{\varphi\in\Phi}\Sigma_{\psi\in\Psi}|\int_A \psi \cdot \varphi \cdot f|$. This absolute convergence justifies interchanging the order of summation in the above equation; the resulting double sum clearly equals $\Sigma_{\psi\in\Psi}\int_A \psi \cdot f$. Finally, this result applied to $|f|$ proves convergence of $\Sigma_{\psi\in\Psi}\int_A \psi \cdot |f|$.

(2) If A is contained in the closed rectangle B and $|f(x)| \leq M$ for $x \in A$, and $F \subset \Phi$ is finite, then

$$\sum_{\varphi \in F} \int_A \varphi \cdot |f| \leq \sum_{\varphi \in F} M \int_A \varphi = M \int_A \sum_{\varphi \in F} \varphi \leq Mv(B),$$

since $\Sigma_{\varphi \in F}\, \varphi \leq 1$ on A.

(3) If $\varepsilon > 0$ there is (Problem 3-22) a compact Jordan-measurable $C \subset A$ such that $\int_{A-C} 1 < \varepsilon$. There are only finitely many $\varphi \in \Phi$ which are non-zero on C. If $F \subset \Phi$ is any finite collection which includes these, and $\int_A f$ has its old meaning, then

$$\begin{aligned} \left| \int_A f - \sum_{\varphi \in F} \int_A \varphi \cdot f \right| &\leq \int_A \left| f - \sum_{\varphi \in F} \varphi \cdot f \right| \\ &\leq M \int_A \left(1 - \sum_{\varphi \in F} \varphi\right) \\ &= M \int_A \sum_{\varphi \in \Phi - F} \varphi \leq M \int_{A-C} 1 \leq M\varepsilon. \quad \blacksquare \end{aligned}$$

Problems. **3-37.** (a) Suppose that $f\colon (0,1) \to \mathbf{R}$ is a non-negative continuous function. Show that $\int_{(0,1)} f$ exists if and only if $\lim_{\varepsilon \to 0} \int_{\varepsilon}^{1-\varepsilon} f$ exists.

(b) Let $A_n = [1 - 1/2^n, 1 - 1/2^{n+1}]$. Suppose that $f\colon (0,1) \to \mathbf{R}$ satisfies $\int_{A_n} f = (-1)^n/n$ and $f(x) = 0$ for $x \notin$ any A_n. Show that $\int_{(0,1)} f$ does not exist, but $\lim_{\varepsilon \to 0} \int_{(\varepsilon, 1-\varepsilon)} f = \log 2$.

3-38. Let A_n be a closed set contained in $(n, n+1)$. Suppose that $f\colon \mathbf{R} \to \mathbf{R}$ satisfies $\int_{A_n} f = (-1)^n/n$ and $f = 0$ for $x \notin$ any A_n. Find two partitions of unity Φ and Ψ such that $\Sigma_{\varphi \in \Phi} \int_{\mathbf{R}} \varphi \cdot f$ and $\Sigma_{\psi \in \Psi} \int_{\mathbf{R}} \psi \cdot f$ converge absolutely to different values.

CHANGE OF VARIABLE

If $g\colon [a,b] \to \mathbf{R}$ is continuously differentiable and $f\colon \mathbf{R} \to \mathbf{R}$ is continuous, then, as is well known,

$$\int_{g(a)}^{g(b)} f = \int_a^b (f \circ g) \cdot g'.$$

The proof is very simple: if $F' = f$, then $(F \circ g)' = (f \circ g) \cdot g'$; thus the left side is $F(g(b)) - F(g(a))$, while the right side is $F \circ g(b) - F \circ g(a) = F(g(b)) - F(g(a))$.

We leave it to the reader to show that if g is 1-1, then the above formula can be written

$$\int_{g((a,b))} f = \int_{(a,b)} f \circ g \cdot |g'|.$$

(Consider separately the cases where g is increasing and where g is decreasing.) The generalization of this formula to higher dimensions is by no means so trivial.

3-13 Theorem. *Let $A \subset \mathbf{R}^n$ be an open set and $g\colon A \to \mathbf{R}^n$ a 1-1, continuously differentiable function such that $\det g'(x) \neq 0$ for all $x \in A$. If $f\colon g(A) \to \mathbf{R}$ is integrable, then*

$$\int_{g(A)} f = \int_A (f \circ g)|\det g'|.$$

Proof. We begin with some important reductions.

1. Suppose there is an admissible cover $\mathcal{O}$ for A such that for each $U \in \mathcal{O}$ and any integrable f we have

$$\int_{g(U)} f = \int_U (f \circ g)|\det g'|.$$

Then the theorem is true for all of A. (Since g is automatically 1-1 in an open set around each point, it is not surprising that this is the only part of the proof using the fact that g is 1-1 on all of A.)

Proof of (1). The collection of all $g(U)$ is an open cover of $g(A)$. Let Φ be a partition of unity subordinate to this cover. If $\varphi = 0$ outside of $g(U)$, then, since g is 1-1, we have $(\varphi \cdot f) \circ g$

$= 0$ outside of U. Therefore the equation

$$\int_{g(U)} \varphi \cdot f = \int_U [(\varphi \cdot f) \circ g] |\det g'|.$$

can be written

$$\int_{g(A)} \varphi \cdot f = \int_A [(\varphi \cdot f) \circ g] |\det g'|.$$

Hence

$$\begin{aligned}
\int_{g(A)} f = \sum_{\varphi \in \Phi} \int_{g(A)} \varphi \cdot f &= \sum_{\varphi \in \Phi} \int_A [(\varphi \cdot f) \circ g] |\det g'| \\
&= \sum_{\varphi \in \Phi} \int_A (\varphi \circ g)(f \circ g) |\det g'| \\
&= \int_A (f \circ g) |\det g'|.
\end{aligned}$$

Remark. The theorem also follows from the assumption that

$$\int_V f = \int_{g^{-1}(V)} (f \circ g) |\det g'|$$

for V in some admissible cover of $g(A)$. This follows from (1) applied to g^{-1}.

2. It suffices to prove the theorem for the function $f = 1$.

Proof of (2). If the theorem holds for $f = 1$, it holds for constant functions. Let V be a rectangle in $g(A)$ and P a partition of V. For each subrectangle S of P let f_S be the constant function $m_S(f)$. Then

$$\begin{aligned}
L(f,P) &= \sum_S m_S(f) \cdot v(S) = \sum_S \int_{\text{int } S} f_S \\
&= \sum_S \int_{g^{-1}(\text{int } S)} (f_S \circ g) |\det g' \leq \sum_S \int_{g^{-1}(\text{int } S)} (f \circ g) |\det g'| \\
&\leq \int_{g^{-1}(V)} (f \circ g) |\det g'|.
\end{aligned}$$

Since $\int_V f$ is the least upper bound of all $L(f,P)$, this proves that $\int_V f \leq \int_{g^{-1}(V)} (f \circ g) |\det g'|$. A similar argument, letting $f_S = M_S(f)$, shows that $\int_V f \geq \int_{g^{-1}(V)} (f \circ g) |\det g'|$. The result now follows from the above Remark.

3. If the theorem is true for $g\colon A \to \mathbf{R}^n$ and for $h\colon B \to \mathbf{R}^n$, where $g(A) \subset B$, then it is true for $h \circ g\colon A \to \mathbf{R}^n$.

Proof of (3).

$$\int_{h \circ g(A)} f = \int_{h(g(A))} f = \int_{g(A)} (f \circ h)|\det h'|$$

$$= \int_A [(f \circ h) \circ g] \cdot [|\det h'| \circ g] \cdot |\det g'|$$

$$= \int_A f \circ (h \circ g)|\det (h \circ g)'|.$$

4. The theorem is true if g is a linear transformation.

Proof of (4). By (1) and (2) it suffices to show for any open rectangle U that

$$\int_{g(U)} 1 = \int_U |\det g'|.$$

This is Problem 3-35.

Observations (3) and (4) together show that we may assume for any particular $a \in A$ that $g'(a)$ is the identity matrix: in fact, if T is the linear transformation $Dg(a)$, then $(T^{-1} \circ g)'(a) = I$; since the theorem is true for T, if it is true for $T^{-1} \circ g$ it will be true for g.

We are now prepared to give the proof, which preceeds by induction on n. The remarks before the statement of the theorem, together with (1) and (2), prove the case $n = 1$. Assuming the theorem in dimension $n - 1$, we prove it in dimension n. For each $a \in A$ we need only find an open set U with $a \in U \subset A$ for which the theorem is true. Moreover we may assume that $g'(a) = I$.

Define $h\colon A \to \mathbf{R}^n$ by $h(x) = (g^1(x), \ldots ,g^{n-1}(x),x^n)$. Then $h'(a) = I$. Hence in some open U' with $a \in U' \subset A$, the function h is 1-1 and $\det h'(x) \neq 0$. We can thus define $k\colon h(U') \to \mathbf{R}^n$ by $k(x) = (x^1, \ldots ,x^{n-1},g^n(h^{-1}(x)))$ and $g = k \circ h$. We have thus expressed g as the composition

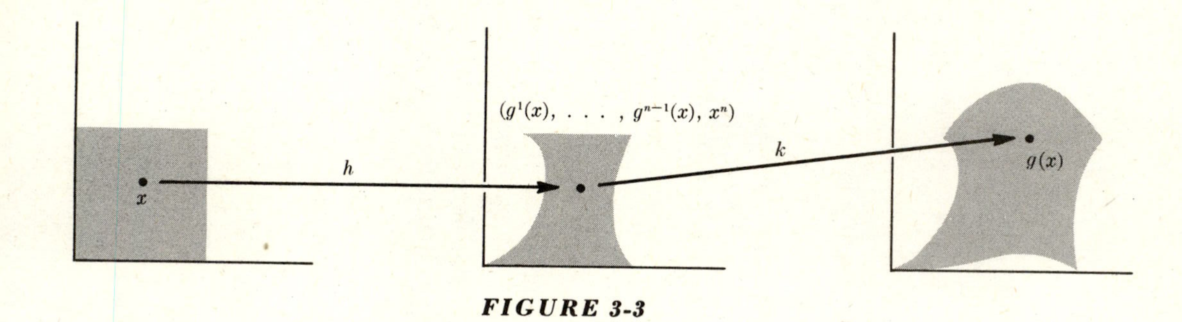

FIGURE 3-3

of two maps, each of which changes fewer than n coordinates (Figure 3-3).

We must attend to a few details to ensure that k is a function of the proper sort. Since

$$(g^n \circ h^{-1})'(h(a)) = (g^n)'(a) \cdot [h'(a)]^{-1} = (g^n)'(a),$$

we have $D_n(g^n \circ h^{-1})(h(a)) = D_n g^n(a) = 1$, so that $k'(h(a)) = I$. Thus in some open set V with $h(a) \in V \subset h(U')$, the function k is 1-1 and $\det k'(x) \neq 0$. Letting $U = k^{-1}(V)$ we now have $g = k \circ h$, where $h\colon U \to \mathbf{R}^n$ and $k\colon V \to \mathbf{R}^n$ and $h(U) \subset V$. By (3) it suffices to prove the theorem for h and k. We give the proof for h; the proof for k is similar and easier.

Let $W \subset U$ be a rectangle of the form $D \times [a_n,b_n]$, where D is a rectangle in $\mathbf{R}^{n-1}$. By Fubini's theorem

$$\int_{h(W)} 1 = \int_{[a_n,b_n]} \left(\int_{h(D \times \{x^n\})} 1\, dx^1 \cdots dx^{n-1} \right) dx^n.$$

Let $h_{x^n}\colon D \to \mathbf{R}^{n-1}$ be defined by $h_{x^n}(x^1, \ldots ,x^{n-1}) = (g^1(x^1, \ldots ,x^n), \ldots ,g^{n-1}(x^1, \ldots ,x^n))$. Then each h_{x^n} is clearly 1-1 and

$$\det (h_{x^n})'(x^1, \ldots ,x^{n-1}) = \det h'(x^1, \ldots ,x^n) \neq 0.$$

Moreover

$$\int_{h(D \times \{x^n\})} 1\, dx^1 \cdots dx^{n-1} = \int_{h_{x^n}(D)} 1\, dx^1 \cdots dx^{n-1}.$$

Applying the theorem in the case $n - 1$ therefore gives

$$\begin{aligned}
\int_{h(W)} 1 &= \int_{[a_n,b_n]} \left(\int_{h_{x^n}(D)} 1\, dx^1 \cdots dx^{n-1} \right) dx^n \\
&= \int_{[a_n,b_n]} \left(\int_D |\det(h_{x^n})'(x^1, \ldots ,x^{n-1})|\, dx^1 \cdots dx^{n-1} \right) dx^n \\
&= \int_{[a_n,b_n]} \left(\int_D |\det h'(x^1, \ldots ,x^n)|\, dx^1 \cdots dx^{n-1} \right) dx^n \\
&= \int_W |\det h'|. \quad \blacksquare
\end{aligned}$$

The condition $\det g'(x) \neq 0$ may be eliminated from the hypotheses of Theorem 3-13 by using the following theorem, which often plays an unexpected role.

3-14. Theorem (Sard's Theorem). *Let $g: A \to \mathbf{R}^n$ be continuously differentiable, where $A \subset \mathbf{R}^n$ is open, and let $B = \{x \in A: \det g'(x) = 0\}$. Then $g(B)$ has measure* 0.

Proof. Let $U \subset A$ be a closed rectangle such that all sides of U have length l, say. Let $\varepsilon > 0$. If N is sufficiently large and U is divided into N^n rectangles, with sides of length l/N, then for each of these rectangles S, if $x \in S$ we have

$$|Dg(x)(y - x) - g(y) - g(x)| < \varepsilon|x - y| \leq \varepsilon \sqrt{n}\,(l/N)$$

for all $y \in S$. If S intersects B we can choose $x \in S \cap B$; since $\det g'(x) = 0$, the set $\{Dg(x)(y - x): y \in S\}$ lies in an $(n - 1)$-dimensional subspace V of $\mathbf{R}^n$. Therefore the set $\{g(y) - g(x): y \in S\}$ lies within $\varepsilon \sqrt{n}\,(l/N)$ of V, so that $\{g(y): y \in S\}$ lies within $\varepsilon \sqrt{n}\,(l/N)$ of the $(n - 1)$-plane $V + g(x)$. On the other hand, by Lemma 2-10 there is a number M such that

$$|g(x) - g(y)| < M|x - y| \leq M \sqrt{n}\,(l/N).$$

Thus, if S intersects B, the set $\{g(y): y \in S\}$ is contained in a cylinder whose height is $<2\varepsilon \sqrt{n}\,(l/N)$ and whose base is an $(n - 1)$-dimensional sphere of radius $<M \sqrt{n}\,(l/N)$. This cylinder has volume $<C(l/N)^n\varepsilon$ for some constant C. There are at most N^n such rectangles S, so $g(U \cap B)$ lies in a set of volume $<C(l/N)^n \cdot \varepsilon \cdot N^n = Cl^n \cdot \varepsilon$. Since this is true for all $\varepsilon > 0$, the set $g(U \cap B)$ has measure 0. Since (Problem 3-13) we can cover all of A with a sequence of such rectangles U, the desired result follows from Theorem 3-4. ▮

Theorem 3-14 is actually only the easy part of Sard's Theorem. The statement and proof of the deeper result will be found in [17], page 47.

Problems. **3-39.** Use Theorem 3-14 to prove Theorem 3-13 without the assumption $\det g'(x) \neq 0$.

3-40. If $g: \mathbf{R}^n \to \mathbf{R}^n$ and $\det g'(x) \neq 0$, prove that in some open set containing x we can write $g = T \circ g_n \circ \cdots \circ g_1$, where g_i is of the form $g_i(x) = (x^1, \ldots, f_i(x), \ldots, x^n)$, and T is a linear transformation. Show that we can write $g = g_n \circ \cdots \circ g_1$ if and only if $g'(x)$ is a diagonal matrix.

3-41. Define $f: \{r: r > 0\} \times (0,2\pi) \to \mathbf{R}^2$ by $f(r,\theta) = (r \cos \theta, r \sin \theta)$.

(a) Show that f is 1-1, compute $f'(r,\theta)$, and show that $\det f'(r,\theta) \neq 0$ for all (r,θ). Show that $f(\{r: r > 0\} \times (0,2\pi))$ is the set A of Problem 2-23.

(b) If $P = f^{-1}$, show that $P(x,y) = (r(x,y),\theta(x,y))$, where

$$r(x,y) = \sqrt{x^2 + y^2},$$

$$\theta(x,y) = \begin{cases} \arctan y/x & x > 0,\ y > 0, \\ \pi + \arctan y/x & x < 0, \\ 2\pi + \arctan y/x & x > 0,\ y < 0, \\ \pi/2 & x = 0,\ y > 0, \\ 3\pi/2 & x = 0,\ y < 0. \end{cases}$$

(Here arctan denotes the inverse of the function $\tan: (-\pi/2,\pi/2) \to \mathbf{R}$.) Find $P'(x,y)$. The function P is called the **polar coordinate system** on A.

(c) Let $C \subset A$ be the region between the circles of radii r_1 and r_2 and the half-lines through 0 which make angles of θ_1 and θ_2 with the x-axis. If $h: C \to \mathbf{R}$ is integrable and $h(x,y) = g(r(x,y),\theta(x,y))$, show that

$$\int_C h = \int_{r_1}^{r_2} \int_{\theta_1}^{\theta_2} rg(r,\theta)d\theta\, dr.$$

If $B_r = \{(x,y): x^2 + y^2 \leq r^2\}$, show that

$$\int_{B_r} h = \int_0^r \int_0^{2\pi} rg(r,\theta)d\theta\, dr.$$

(d) If $C_r = [-r,r] \times [-r,r]$, show that

$$\int_{B_r} e^{-(x^2+y^2)}\, dx\, dy = \pi(1 - e^{-r^2})$$

and

$$\int_{C_r} e^{-(x^2+y^2)}\, dx\, dy = \left(\int_{-r}^{r} e^{-x^2}\, dx\right)^2.$$

(e) Prove that

$$\lim_{r\to\infty} \int_{B_r} e^{-(x^2+y^2)}\, dx\, dy = \lim_{r\to\infty} \int_{C_r} e^{-(x^2+y^2)}\, dx\, dy$$

and conclude that

$$\int_{-\infty}^{\infty} e^{-x^2}\,dx = \sqrt{\pi}.$$

"A mathematician is one to whom *that* is as obvious as that twice two makes four is to you. Liouville was a mathematician."

—LORD KELVIN

4

Integration on Chains

ALGEBRAIC PRELIMINARIES

If V is a vector space (over $\mathbf{R}$), we will denote the k-fold product $V \times \cdots \times V$ by V^k. A function $T\colon V^k \to \mathbf{R}$ is called **multilinear** if for each i with $1 \leq i \leq k$ we have

$$\begin{aligned} T(v_1, \ldots, v_i + v_i', \ldots, v_k) &= T(v_1, \ldots, v_i, \ldots, v_k) \\ &\quad + T(v_1, \ldots, v_i', \ldots, v_k), \\ T(v_1, \ldots, av_i, \ldots, v_k) &= aT(v_1, \ldots, v_i, \ldots, v_k). \end{aligned}$$

A multilinear function $T\colon V^k \to \mathbf{R}$ is called a $\boldsymbol{k}$**-tensor** on V and the set of all k-tensors, denoted $\mathfrak{J}^k(V)$, becomes a vector space (over $\mathbf{R}$) if for $S,T \in \mathfrak{J}^k(V)$ and $a \in \mathbf{R}$ we define

$$\begin{aligned} (S + T)(v_1, \ldots, v_k) &= S(v_1, \ldots, v_k) + T(v_1, \ldots, v_k), \\ (aS)(v_1, \ldots, v_k) &= a \cdot S(v_1, \ldots, v_k). \end{aligned}$$

There is also an operation connecting the various spaces $\mathfrak{J}^k(V)$. If $S \in \mathfrak{J}^k(V)$ and $T \in \mathfrak{J}^l(V)$, we define the **tensor product** $S \otimes T \in \mathfrak{J}^{k+l}(V)$ by

$$\begin{aligned} S \otimes T(v_1, \ldots, v_k, v_{k+1}, \ldots, v_{k+l}) \\ = S(v_1, \ldots, v_k) \cdot T(v_{k+1}, \ldots, v_{k+l}). \end{aligned}$$

Note that the order of the factors S and T is crucial here since $S \otimes T$ and $T \otimes S$ are far from equal. The following properties of $\otimes$ are left as easy exercises for the reader.

$$\begin{aligned}(S_1 + S_2) \otimes T &= S_1 \otimes T + S_2 \otimes T,\\ S \otimes (T_1 + T_2) &= S \otimes T_1 + S \otimes T_2,\\ (aS) \otimes T &= S \otimes (aT) = a(S \otimes T),\\ (S \otimes T) \otimes U &= S \otimes (T \otimes U).\end{aligned}$$

Both $(S \otimes T) \otimes U$ and $S \otimes (T \otimes U)$ are usually denoted simply $S \otimes T \otimes U$; higher-order products $T_1 \otimes \cdots \otimes T_r$ are defined similarly.

The reader has probably already noticed that $\mathfrak{J}^1(V)$ is just the dual space V^*. The operation $\otimes$ allows us to express the other vector spaces $\mathfrak{J}^k(V)$ in terms of $\mathfrak{J}^1(V)$.

4-1 Theorem. *Let $v_1, \ldots, v_n$ be a basis for V, and let $\varphi_1, \ldots, \varphi_n$ be the dual basis, $\varphi_i(v_j) = \delta_{ij}$. Then the set of all k-fold tensor products*

$$\varphi_{i_1} \otimes \cdots \otimes \varphi_{i_k} \qquad 1 \leq i_1, \ldots, i_k \leq n$$

is a basis for $\mathfrak{J}^k(V)$, which therefore has dimension n^k.

Proof. Note that

$$\begin{aligned}\varphi_{i_1} \otimes \cdots \otimes \varphi_{i_k}(v_{j_1}, \ldots, v_{j_k}) &= \delta_{i_1,j_1} \cdot \ldots \cdot \delta_{i_k,j_k}\\ &= \begin{cases} 1 & \text{if } j_1 = i_1, \ldots, j_k = i_k,\\ 0 & \text{otherwise.}\end{cases}\end{aligned}$$

If $w_1, \ldots, w_k$ are k vectors with $w_i = \sum_{j=1}^n a_{ij} v_j$ and T is in $\mathfrak{J}^k(V)$, then

$$\begin{aligned}T(w_1, \ldots, w_k) &= \sum_{j_1, \ldots, j_k = 1}^n a_{1,j_1} \cdot \ldots \cdot a_{k,j_k} T(v_{j_1}, \ldots, v_{j_k})\\ &= \sum_{i_1, \ldots, i_k = 1}^n T(v_{i_1}, \ldots, v_{i_k}) \cdot \varphi_{i_1} \otimes \cdots \otimes \varphi_{i_k}(w_1, \ldots, w_k).\end{aligned}$$

Thus $T = \sum_{i_1, \ldots, i_k = 1}^n T(v_{i_1}, \ldots, v_{i_k}) \cdot \varphi_{i_1} \otimes \cdots \otimes \varphi_{i_k}$.

Consequently the $\varphi_{i_1} \otimes \cdots \otimes \varphi_{i_k}$ span $\mathfrak{J}^k(V)$.

Suppose now that there are numbers $a_{i_1,\ldots,i_k}$ such that

$$\sum_{i_1,\ldots,i_k=1}^{n} a_{i_1,\ldots,i_k} \cdot \varphi_{i_1} \otimes \cdots \otimes \varphi_{i_k} = 0.$$

Applying both sides of this equation to $(v_{j_1},\ldots,v_{j_k})$ yields $a_{j_1,\ldots,j_k} = 0$. Thus the $\varphi_{i_1} \otimes \cdots \otimes \varphi_{i_k}$ are linearly independent. ▌

One important construction, familiar for the case of dual spaces, can also be made for tensors. If $f\colon V \to W$ is a linear transformation, a linear transformation $f^*\colon \mathfrak{J}^k(W) \to \mathfrak{J}^k(V)$ is defined by

$$f^*T(v_1,\ldots,v_k) = T(f(v_1),\ldots,f(v_k))$$

for $T \in \mathfrak{J}^k(W)$ and $v_1,\ldots,v_k \in V$. It is easy to verify that $f^*(S \otimes T) = f^*S \otimes f^*T$.

The reader is already familiar with certain tensors, aside from members of V^*. The first example is the inner product $\langle,\rangle \in \mathfrak{J}^2(\mathbf{R}^n)$. On the grounds that any good mathematical commodity is worth generalizing, we define an **inner product** on V to be a 2-tensor T such that T is **symmetric,** that is $T(v,w) = T(w,v)$ for $v,w \in V$ and such that T is **positive-definite,** that is, $T(v,v) > 0$ if $v \neq 0$. We distinguish $\langle,\rangle$ as the **usual inner product** on $\mathbf{R}^n$. The following theorem shows that our generalization is not too general.

4-2 Theorem. *If T is an inner product on V, there is a basis $v_1,\ldots,v_n$ for V such that $T(v_i,v_j) = \delta_{ij}$. (Such a basis is called* **orthonormal** *with respect to T.) Consequently there is an isomorphism $f\colon \mathbf{R}^n \to V$ such that $T(f(x),f(y)) = \langle x,y\rangle$ for $x,y \in \mathbf{R}^n$. In other words $f^*T = \langle,\rangle$.*

Proof. Let $w_1,\ldots,w_n$ be any basis for V. Define

$$w_1' = w_1,$$

$$w_2' = w_2 - \frac{T(w_1',w_2)}{T(w_1',w_1')} \cdot w_1',$$

$$w_3' = w_3 - \frac{T(w_1',w_3)}{T(w_1',w_1')} \cdot w_1' - \frac{T(w_2',w_3)}{T(w_2',w_2')} \cdot w_2',$$

etc.

It is easy to check that $T(w_i',w_j') = 0$ if $i \neq j$ and $w_i' \neq 0$ so that $T(w_i',w_i') > 0$. Now define $v_i = w_i'/\sqrt{T(w_i',w_i')}$. The isomorphism f may be defined by $f(e_i) = v_i$. ∎

Despite its importance, the inner product plays a far lesser role than another familiar, seemingly ubiquitous function, the tensor $\det \in \mathfrak{I}^n(\mathbf{R}^n)$. In attempting to generalize this function, we recall that interchanging two rows of a matrix changes the sign of its determinant. This suggests the following definition. A k-tensor $\omega \in \mathfrak{I}^k(V)$ is called **alternating** if

$$\omega(v_1, \ldots ,v_i, \ldots ,v_j, \ldots ,v_k) = -\omega(v_1, \ldots ,v_j, \ldots ,v_i, \ldots ,v_k) \quad \text{for all } v_1, \ldots ,v_k \in V.$$

(In this equation v_i and v_j are interchanged and all other v's are left fixed.) The set of all alternating k-tensors is clearly a subspace $\Lambda^k(V)$ of $\mathfrak{I}^k(V)$. Since it requires considerable work to produce the determinant, it is not surprising that alternating k-tensors are difficult to write down. There is, however, a uniform way of expressing all of them. Recall that the sign of a permutation σ, denoted $\operatorname{sgn} \sigma$, is $+1$ if σ is even and -1 if σ is odd. If $T \in \mathfrak{I}^k(V)$, we define $\operatorname{Alt}(T)$ by

$$\operatorname{Alt}(T)(v_1, \ldots ,v_k) = \frac{1}{k!} \sum_{\sigma \in S_k} \operatorname{sgn} \sigma \cdot T(v_{\sigma(1)}, \ldots ,v_{\sigma(k)}),$$

where S_k is the set of all permutations of the numbers 1 to k.

4-3 Theorem

(1) *If $T \in \mathfrak{I}^k(V)$, then $Alt(T) \in \Lambda^k(V)$.*
(2) *If $\omega \in \Lambda^k(V)$, then $Alt(\omega) = \omega$.*
(3) *If $T \in \mathfrak{I}^k(V)$, then $Alt(Alt(T)) = Alt(T)$.*

Proof

(1) Let (i,j) be the permutation that interchanges i and j and leaves all other numbers fixed. If $\sigma \in S_k$, let $\sigma' = \sigma \cdot (i,j)$. Then

$$\begin{aligned}
&\text{Alt}(T)(v_1, \ldots, v_j, \ldots, v_i, \ldots, v_k) \\
&= \frac{1}{k!} \sum_{\sigma \in S_k} \text{sgn}\, \sigma \cdot T(v_{\sigma(1)}, \ldots, v_{\sigma(j)}, \ldots, v_{\sigma(i)}, \ldots, v_{\sigma(k)}) \\
&= \frac{1}{k!} \sum_{\sigma \in S_k} \text{sgn}\, \sigma \cdot T(v_{\sigma'(1)}, \ldots, v_{\sigma'(i)}, \ldots, v_{\sigma'(j)}, \ldots, v_{\sigma'(k)}) \\
&= \frac{1}{k!} \sum_{\sigma' \in S_k} -\text{sgn}\, \sigma' \cdot T(v_{\sigma'(1)}, \ldots, v_{\sigma'(k)}) \\
&= -\text{Alt}(T)(v_1, \ldots, v_k).
\end{aligned}$$

(2) If $\omega \in \Lambda^k(V)$, and $\sigma = (i,j)$, then $\omega(v_{\sigma(1)}, \ldots, v_{\sigma(k)}) = \text{sgn}\, \sigma \cdot \omega(v_1, \ldots, v_k)$. Since every σ is a product of permutations of the form (i,j), this equation holds of all σ. Therefore

$$\begin{aligned}
\text{Alt}(\omega)(v_1, \ldots, v_k) &= \frac{1}{k!} \sum_{\sigma \in S_k} \text{sgn}\, \sigma \cdot \omega(v_{\sigma(1)}, \ldots, v_{\sigma(k)}) \\
&= \frac{1}{k!} \sum_{\sigma \in S_k} \text{sgn}\, \sigma \cdot \text{sgn}\, \sigma \cdot \omega(v_1, \ldots, v_k) \\
&= \omega(v_1, \ldots, v_k).
\end{aligned}$$

(3) follows immediately from (1) and (2). ▮

To determine the dimensions of $\Lambda^k(V)$, we would like a theorem analogous to Theorem 4-1. Of course, if $\omega \in \Lambda^k(V)$ and $\eta \in \Lambda^l(V)$, then $\omega \otimes \eta$ is usually not in $\Lambda^{k+l}(V)$. We will therefore define a new product, the **wedge** product $\omega \wedge \eta \in \Lambda^{k+l}(V)$ by

$$\omega \wedge \eta = \frac{(k+l)!}{k!\, l!} \text{Alt}(\omega \otimes \eta).$$

(The reason for the strange coefficient will appear later.) The following properties of $\wedge$ are left as an exercise for the reader:

$$\begin{aligned}
(\omega_1 + \omega_2) \wedge \eta &= \omega_1 \wedge \eta + \omega_2 \wedge \eta, \\
\omega \wedge (\eta_1 + \eta_2) &= \omega \wedge \eta_1 + \omega \wedge \eta_2, \\
a\omega \wedge \eta = \omega \wedge a\eta &= a(\omega \wedge \eta), \\
\omega \wedge \eta &= (-1)^{kl} \eta \wedge \omega, \\
f^*(\omega \wedge \eta) &= f^*(\omega) \wedge f^*(\eta).
\end{aligned}$$

The equation $(\omega \wedge \eta) \wedge \theta = \omega \wedge (\eta \wedge \theta)$ is true but requires more work.

4-4 Theorem

(1) *If* $S \in \mathfrak{J}^k(V)$ and $T \in \mathfrak{J}^l(V)$ *and* $Alt(S) = 0$, *then*

$$Alt(S \otimes T) = Alt(T \otimes S) = 0.$$

(2) $$Alt(Alt(\omega \otimes \eta) \otimes \theta) = Alt(\omega \otimes \eta \otimes \theta) = Alt(\omega \otimes Alt(\eta \otimes \theta)).$$

(3) *If* $\omega \in \Lambda^k(V)$, $\eta \in \Lambda^l(V)$, *and* $\theta \in \Lambda^m(V)$, *then*

$$(\omega \wedge \eta) \wedge \theta = \omega \wedge (\eta \wedge \theta) = \frac{(k+l+m)!}{k!\,l!\,m!} Alt(\omega \otimes \eta \otimes \theta).$$

Proof

(1)

$$(k+l)!\,Alt(S \otimes T)(v_1, \ldots, v_{k+l}) = \sum_{\sigma \in S_{k+l}} \operatorname{sgn} \sigma \cdot S(v_{\sigma(1)}, \ldots, v_{\sigma(k)}) \cdot T(v_{\sigma(k+1)}, \ldots, v_{\sigma(k+l)}).$$

If $G \subset S_{k+l}$ consists of all σ which leave $k+1, \ldots, k+l$ fixed, then

$$\sum_{\sigma \in G} \operatorname{sgn} \sigma \cdot S(v_{\sigma(1)}, \ldots, v_{\sigma(k)}) \cdot T(v_{\sigma(k+1)}, \ldots, v_{\sigma(k+l)}) = \left[\sum_{\sigma' \in S_k} \operatorname{sgn} \sigma' \cdot S(v_{\sigma'(1)}, \ldots, v_{\sigma'(k)}) \right] \cdot T(v_{k+1}, \ldots, v_{k+l}) = 0.$$

Suppose now that $\sigma_0 \notin G$. Let $G \cdot \sigma_0 = \{\sigma \cdot \sigma_0 \colon \sigma \in G\}$ and let $v_{\sigma_0(1)}, \ldots, v_{\sigma_0(k+l)} = w_1, \ldots, w_{k+l}$. Then

$$\sum_{\sigma \in G \cdot \sigma_0} \operatorname{sgn} \sigma \cdot S(v_{\sigma(1)}, \ldots, v_{\sigma(k)}) \cdot T(v_{\sigma(k+1)}, \ldots, v_{\sigma(k+l)}) = \left[\operatorname{sgn} \sigma_0 \cdot \sum_{\sigma' \in G} \operatorname{sgn} \sigma' \cdot S(w_{\sigma'(1)}, \ldots, w_{\sigma'(k)}) \right] \cdot T(w_{k+1}, \ldots, w_{k+l}) = 0.$$

Notice that $G \cap G \cdot \sigma_0 = \emptyset$. In fact, if $\sigma \in G \cap G \cdot \sigma_0$, then $\sigma = \sigma' \cdot \sigma_0$ for some $\sigma' \in G$ and $\sigma_0 = \sigma \cdot (\sigma')^{-1} \in G$, a contradiction. We can then continue in this way, breaking S_{k+l} up into disjoint subsets; the sum over each subset is 0, so that the sum over S_{k+l} is 0. The relation $\text{Alt}(T \otimes S) = 0$ is proved similarly.

(2) We have

$$\text{Alt}(\text{Alt}(\eta \otimes \theta) - \eta \otimes \theta) = \text{Alt}(\eta \otimes \theta) - \text{Alt}(\eta \otimes \theta) = 0.$$

Hence by (1) we have

$$\begin{aligned} 0 &= \text{Alt}(\omega \otimes [\text{Alt}(\eta \otimes \theta) - \eta \otimes \theta]) \\ &= \text{Alt}(\omega \otimes \text{Alt}(\eta \otimes \theta)) - \text{Alt}(\omega \otimes \eta \otimes \theta). \end{aligned}$$

The other equality is proved similarly.

(3)
$$\begin{aligned} (\omega \wedge \eta) \wedge \theta &= \frac{(k+l+m)!}{(k+l)!\,m!} \text{Alt}((\omega \wedge \eta) \otimes \theta) \\ &= \frac{(k+l+m)!}{(k+l)!\,m!} \frac{(k+l)!}{k!\,l!} \text{Alt}(\omega \otimes \eta \otimes \theta). \end{aligned}$$

The other equality is proved similarly. ▮

Naturally $\omega \wedge (\eta \wedge \theta)$ and $(\omega \wedge \eta) \wedge \theta$ are both denoted simply $\omega \wedge \eta \wedge \theta$, and higher-order products $\omega_1 \wedge \cdots \wedge \omega_r$ are defined similarly. If $v_1, \ldots, v_n$ is a basis for V and $\varphi_1, \ldots, \varphi_n$ is the dual basis, a basis for $\Lambda^k(V)$ can now be constructed quite easily.

4-5 Theorem. *The set of all*

$$\varphi_{i_1} \wedge \cdots \wedge \varphi_{i_k} \qquad 1 \leq i_1 < i_2 < \cdots < i_k \leq n$$

is a basis for $\Lambda^k(V)$, *which therefore has dimension*

$$\binom{n}{k} = \frac{n!}{k!(n-k)!}.$$

Proof. If $\omega \in \Lambda^k(V) \subset \mathfrak{J}^k(V)$, then we can write

$$\omega = \sum_{i_1, \ldots, i_k} a_{i_1, \ldots, i_k} \varphi_{i_1} \otimes \cdots \otimes \varphi_{i_k}.$$

Thus

$$\omega = \text{Alt}(\omega) = \sum_{i_1,\ldots,i_k} a_{i_1,\ldots,i_k}\text{Alt}(\varphi_{i_1} \otimes \cdots \otimes \varphi_{i_k}).$$

Since each $\text{Alt}(\varphi_{i_1} \otimes \cdots \otimes \varphi_{i_k})$ is a constant times one of the $\varphi_{i_1} \wedge \cdots \wedge \varphi_{i_k}$, these elements span $\Lambda^k(V)$. Linear independence is proved as in Theorem 4-1 (cf. Problem 4-1). ▌

If V has dimension n, it follows from Theorem 4-5 that $\Lambda^n(V)$ has dimension 1. Thus all alternating n-tensors on V are multiples of any non-zero one. Since the determinant is an example of such a member of $\Lambda^n(\mathbf{R}^n)$, it is not surprising to find it in the following theorem.

4-6 Theorem. *Let $v_1, \ldots, v_n$ be a basis for V, and let $\omega \in \Lambda^n(V)$. If $w_i = \sum_{j=1}^n a_{ij}v_j$ are n vectors in V, then*

$$\omega(w_1, \ldots, w_n) = \det(a_{ij}) \cdot \omega(v_1, \ldots, v_n).$$

Proof. Define $\eta \in \mathfrak{J}^n(\mathbf{R}^n)$ by

$$\eta((a_{11}, \ldots, a_{1n}), \ldots, (a_{n1}, \ldots, a_{nn})) = \omega(\Sigma a_{1j}v_j, \ldots, \Sigma a_{nj}v_j).$$

Clearly $\eta \in \Lambda^n(\mathbf{R}^n)$ so $\eta = \lambda \cdot \det$ for some $\lambda \in \mathbf{R}$ and $\lambda = \eta(e_1, \ldots, e_n) = \omega(v_1, \ldots, v_n)$. ▌

Theorem 4-6 shows that a non-zero $\omega \in \Lambda^n(V)$ splits the bases of V into two disjoint groups, those with $\omega(v_1, \ldots, v_n) > 0$ and those for which $\omega(v_1, \ldots, v_n) < 0$; if $v_1, \ldots, v_n$ and $w_1, \ldots, w_n$ are two bases and $A = (a_{ij})$ is defined by $w_i = \Sigma a_{ij}v_j$, then $v_1, \ldots, v_n$ and $w_1, \ldots, w_n$ are in the same group if and only if $\det A > 0$. This criterion is independent of ω and can always be used to divide the bases of V into two disjoint groups. Either of these two groups is called an **orientation** for V. The orientation to which a basis $v_1, \ldots, v_n$ belongs is denoted $[v_1, \ldots, v_n]$ and the other orientation is denoted $-[v_1, \ldots, v_n]$. In $\mathbf{R}^n$ we define the **usual orientation** as $[e_1, \ldots, e_n]$.

The fact that $\dim \Lambda^n(\mathbf{R}^n) = 1$ is probably not new to you, since det is often defined as the unique element $\omega \in \Lambda^n(\mathbf{R}^n)$ such that $\omega(e_1, \ldots, e_n) = 1$. For a general vector space V there is no extra criterion of this sort to distinguish a particular $\omega \in \Lambda^n(V)$. Suppose, however, that an inner product T for V is given. If $v_1, \ldots, v_n$ and $w_1, \ldots, w_n$ are two bases which are orthonormal with respect to T, and the matrix $A = (a_{ij})$ is defined by $w_i = \sum_{j=1}^n a_{ij}v_j$, then

$$\begin{aligned}\delta_{ij} = T(w_i,w_j) &= \sum_{k,l=1}^n a_{ik}a_{jl}T(v_k,v_l) \\ &= \sum_{k=1}^n a_{ik}a_{jk}.\end{aligned}$$

In other words, if $A^{\mathbf{T}}$ denotes the transpose of the matrix A, then we have $A \cdot A^{\mathbf{T}} = I$, so $\det A = \pm 1$. It follows from Theorem 4-6 that if $\omega \in \Lambda^n(V)$ satisfies $\omega(v_1, \ldots, v_n) = \pm 1$, then $\omega(w_1, \ldots, w_n) = \pm 1$. If an orientation μ for V has also been given, it follows that there is a unique $\omega \in \Lambda^n(V)$ such that $\omega(v_1, \ldots, v_n) = 1$ whenever $v_1, \ldots, v_n$ is an orthonormal basis such that $[v_1, \ldots, v_n] = \mu$. This unique ω is called the **volume element** of V, determined by the inner product T and orientation μ. Note that det is the volume element of $\mathbf{R}^n$ determined by the usual inner product and usual orientation, and that $|\det(v_1, \ldots, v_n)|$ is the volume of the parallelipiped spanned by the line segments from 0 to each of $v_1, \ldots, v_n$.

We conclude this section with a construction which we will restrict to $\mathbf{R}^n$. If $v_1, \ldots, v_{n-1} \in \mathbf{R}^n$ and φ is defined by

$$\varphi(w) = \det \begin{pmatrix} v_1 \\ \cdot \\ \cdot \\ \cdot \\ v_{n-1} \\ w \end{pmatrix},$$

then $\varphi \in \Lambda^1(\mathbf{R}^n)$; therefore there is a unique $z \in \mathbf{R}^n$ such that

$$\langle w,z \rangle = \varphi(w) = \det \begin{pmatrix} v_1 \\ \cdot \\ \cdot \\ \cdot \\ v_{n-1} \\ w \end{pmatrix}$$

This z is denoted $v_1 \times \cdots \times v_{n-1}$ and called the **cross product** of $v_1, \ldots, v_{n-1}$. The following properties are immediate from the definition:

$$v_{\sigma(1)} \times \cdots \times v_{\sigma(n-1)} = \operatorname{sgn} \sigma \cdot v_1 \times \cdots \times v_{n-1},$$
$$v_1 \times \cdots \times av_i \times \cdots \times v_{n-1} = a \cdot (v_1 \times \cdots \times v_{n-1}),$$
$$\begin{aligned} v_1 \times \cdots \times (v_i + v_i') \times \cdots \times v_{n-1} \\ = v_1 \times \cdots \times v_i \times \cdots \times v_{n-1} \\ + v_1 \times \cdots \times v_i' \times \cdots \times v_{n-1}. \end{aligned}$$

It is uncommon in mathematics to have a "product" that depends on more than two factors. In the case of two vectors $v,w \in \mathbf{R}^3$, we obtain a more conventional looking product, $v \times w \in \mathbf{R}^3$. For this reason it is sometimes maintained that the cross product can be defined only in $\mathbf{R}^3$.

Problems. **4-1.*** Let $e_1, \ldots, e_n$ be the usual basis of $\mathbf{R}^n$ and let $\varphi_1, \ldots, \varphi_n$ be the dual basis.

(a) Show that $\varphi_{i_1} \wedge \cdots \wedge \varphi_{i_k} (e_{i_1}, \ldots, e_{i_k}) = 1$. What would the right side be if the factor $(k + l)!/k!l!$ did not appear in the definition of $\wedge$?

(b) Show that $\varphi_{i_1} \wedge \cdots \wedge \varphi_{i_k}(v_1, \ldots, v_k)$ is the determinant of the $k \times k$ minor of $\begin{pmatrix} v_1 \\ \cdot \\ \cdot \\ \cdot \\ v_k \end{pmatrix}$ obtained by selecting columns $i_1, \ldots, i_k$.

4-2. If f: $V \to V$ is a linear transformation and $\dim V = n$, then f^*: $\Lambda^n(V) \to \Lambda^n(V)$ must be multiplication by some constant c. Show that $c = \det f$.

4-3. If $\omega \in \Lambda^n(V)$ is the volume element determined by T and μ, and $w_1, \ldots, w_n \in V$, show that

$$|\omega(w_1, \ldots, w_n)| = \sqrt{\det(g_{ij})},$$

where $g_{ij} = T(w_i, w_j)$. *Hint:* If $v_1, \ldots, v_n$ is an orthonormal basis and $w_i = \sum_{j=1}^n a_{ij} v_j$, show that $g_{ij} = \sum_{k=1}^n a_{ik} a_{kj}$.

4-4. If ω is the volume element of V determined by T and μ, and $f\colon \mathbf{R}^n \to V$ is an isomorphism such that $f^*T = \langle , \rangle$ and such that $[f(e_1), \ldots, f(e_n)] = \mu$, show that $f^*\omega = \det$.

4-5. If $c\colon [0,1] \to (\mathbf{R}^n)^n$ is continuous and each $(c^1(t), \ldots, c^n(t))$ is a basis for $\mathbf{R}^n$, show that $[c^1(0), \ldots, c^n(0)] = [c^1(1), \ldots, c^n(1)]$. *Hint:* Consider $\det \circ c$.

4-6. (a) If $v \in \mathbf{R}^2$, what is $v \times$?

(b) If $v_1, \ldots, v_{n-1} \in \mathbf{R}^n$ are linearly independent, show that $[v_1, \ldots, v_{n-1}, v_1 \times \cdots \times v_{n-1}]$ is the usual orientation of $\mathbf{R}^n$.

4-7. Show that every non-zero $\omega \in \Lambda^n(V)$ is the volume element determined by some inner product T and orientation μ for V.

4-8. If $\omega \in \Lambda^n(V)$ is a volume element, define a "cross product" $v_1 \times \cdots \times v_{n-1}$ in terms of ω.

4-9.* Deduce the following properties of the cross product in $\mathbf{R}^3$:

(a)
$$\begin{array}{lll} e_1 \times e_1 = 0 & e_2 \times e_1 = -e_3 & e_3 \times e_1 = e_2 \\ e_1 \times e_2 = e_3 & e_2 \times e_2 = 0 & e_3 \times e_2 = -e_1 \\ e_1 \times e_3 = -e_2 & e_2 \times e_3 = e_1 & e_3 \times e_3 = 0. \end{array}$$

(b)
$$\begin{aligned} v \times w = (v^2w^3 - v^3w^2)e_1 \\ + (v^3w^1 - v^1w^3)e_2 \\ + (v^1w^2 - v^2w^1)e_3. \end{aligned}$$

(c) $|v \times w| = |v| \cdot |w| \cdot |\sin \theta|$, where $\theta = \angle(v,w)$.
$\langle v \times w, v \rangle = \langle v \times w, w \rangle = 0$.

(d) $\langle v, w \times z \rangle = \langle w, z \times v \rangle = \langle z, v \times w \rangle$
$v \times (w \times z) = \langle v,z \rangle w - \langle v,w \rangle z$
$(v \times w) \times z = \langle v,z \rangle w - \langle w,z \rangle v.$

(e) $|v \times w| = \sqrt{\langle v,v \rangle \cdot \langle w,w \rangle - \langle v,w \rangle^2}$.

4-10. If $w_1, \ldots, w_{n-1} \in \mathbf{R}^n$, show that

$$|w_1 \times \cdots \times w_{n-1}| = \sqrt{\det(g_{ij})},$$

where $g_{ij} = \langle w_i, w_j \rangle$. *Hint:* Apply Problem 4-3 to a certain $(n-1)$-dimensional subspace of $\mathbf{R}^n$.

4-11. If T is an inner product on V, a linear transformation $f\colon V \to V$ is called **self-adjoint** (with respect to T) if $T(x,f(y)) = T(f(x),y)$ for $x,y \in V$. If $v_1, \ldots, v_n$ is an orthonormal basis and $A = (a_{ij})$ is the matrix of f with respect to this basis, show that $a_{ij} = a_{ji}$.

4-12. If $f_1, \ldots, f_{n-1}\colon \mathbf{R}^m \to \mathbf{R}^n$, define $f_1 \times \cdots \times f_{n-1}\colon \mathbf{R}^m \to \mathbf{R}^n$ by $f_1 \times \cdots \times f_{n-1}(p) = f_1(p) \times \cdots \times f_{n-1}(p)$. Use Problem 2-14 to derive a formula for $D(f_1 \times \cdots \times f_{n-1})$ when $f_1, \ldots, f_{n-1}$ are differentiable.

FIELDS AND FORMS

If $p \in \mathbf{R}^n$, the set of all pairs (p,v), for $v \in \mathbf{R}^n$, is denoted $\mathbf{R}^n{}_p$, and called the **tangent space** of $\mathbf{R}^n$ at p. This set is made into a vector space in the most obvious way, by defining

$$(p,v) + (p,w) = (p, v + w),$$
$$a \cdot (p,v) = (p,av).$$

A vector $v \in \mathbf{R}^n$ is often pictured as an arrow from 0 to v; the vector $(p,v) \in \mathbf{R}^n{}_p$ may be pictured (Figure 4-1) as an arrow with the same direction and length, but with initial point p. This arrow goes from p to the point $p + v$, and we therefore

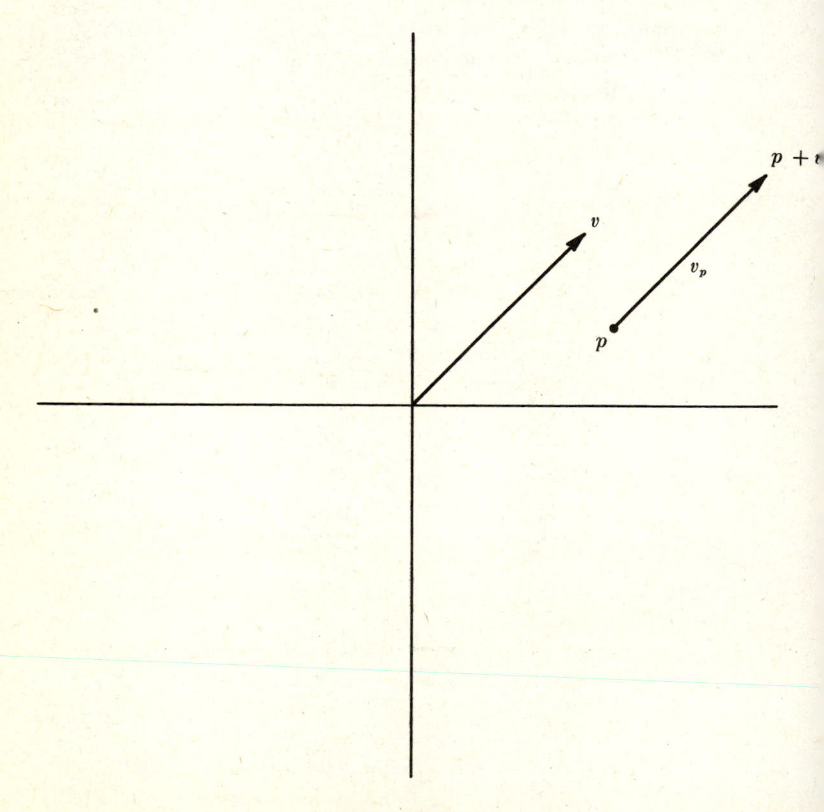

FIGURE 4-1

define $p + v$ to be the **end point** of (p,v). We will usually write (p,v) as v_p (read: the vector v at p).

The vector space $\mathbf{R}^n{}_p$ is so closely allied to $\mathbf{R}^n$ that many of the structures on $\mathbf{R}^n$ have analogues on $\mathbf{R}^n{}_p$. In particular the **usual inner product** $\langle,\rangle_p$ for $\mathbf{R}^n{}_p$ is defined by $\langle v_p,w_p\rangle_p = \langle v,w\rangle$, and the **usual orientation** for $\mathbf{R}^n{}_p$ is $[(e_1)_p, \ldots, (e_n)_p]$.

Any operation which is possible in a vector space may be performed in each $\mathbf{R}^n{}_p$, and most of this section is merely an elaboration of this theme. About the simplest operation in a vector space is the selection of a vector from it. If such a selection is made in each $\mathbf{R}^n{}_p$, we obtain a **vector field** (Figure 4-2). To be precise, a vector field is a function F such that $F(p) \in \mathbf{R}^n{}_p$ for each $p \in \mathbf{R}^n$. For each p there are numbers $F^1(p), \ldots, F^n(p)$ such that

$$F(p) = F^1(p) \cdot (e_1)_p + \cdots + F^n(p) \cdot (e_n)_p.$$

We thus obtain n **component functions** $F^i\colon \mathbf{R}^n \to \mathbf{R}$. The vector field F is called continuous, differentiable, etc., if the functions F^i are. Similar definitions can be made for a vector field defined only on an open subset of $\mathbf{R}^n$. Operations on vectors yield operations on vector fields when applied at each point separately. For example, if F and G are vector fields

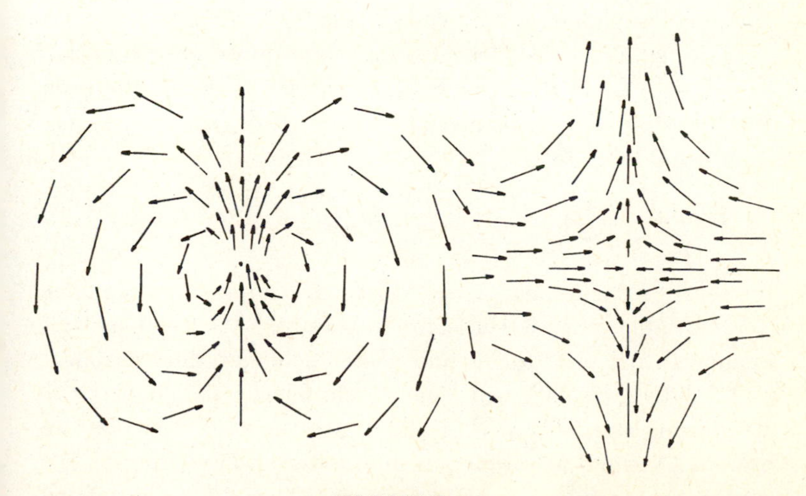

FIGURE 4-2

and f is a function, we define

$$\begin{aligned}(F+G)(p) &= F(p)+G(p),\\ \langle F,G\rangle(p) &= \langle F(p),G(p)\rangle,\\ (f\cdot F)(p) &= f(p)F(p).\end{aligned}$$

If $F_1, \ldots, F_{n-1}$ are vector fields on $\mathbf{R}^n$, then we can similarly define

$$(F_1 \times \cdots \times F_{n-1})(p) = F_1(p) \times \cdots \times F_{n-1}(p).$$

Certain other definitions are standard and useful. We define the **divergence,** $\operatorname{div} F$ of F, as $\sum_{i=1}^n D_iF^i$. If we introduce the formal symbolism

$$\nabla = \sum_{i=1}^n D_i \cdot e_i,$$

we can write, symbolically, $\operatorname{div} F = \langle \nabla, F\rangle$. If $n = 3$ we write, in conformity with this symbolism,

$$\begin{aligned}(\nabla \times F)(p) = (D_2F^3 - D_3F^2)(e_1)_p &\\ + (D_3F^1 - D_1F^3)(e_2)_p &\\ + (D_1F^2 - D_2F^1)(e_3)_p&.\end{aligned}$$

The vector field $\nabla \times F$ is called $\operatorname{curl} F$. The names "divergence" and "curl" are derived from physical considerations which are explained at the end of this book.

Many similar considerations may be applied to a function ω with $\omega(p) \in \Lambda^k(\mathbf{R}^n{}_p)$; such a function is called a ***k*-form** on $\mathbf{R}^n$, or simply a **differential form.** If $\varphi_1(p), \ldots, \varphi_n(p)$ is the dual basis to $(e_1)_p, \ldots, (e_n)_p$, then

$$\omega(p) = \sum_{i_1 < \cdots < i_k} \omega_{i_1,\ldots,i_k}(p) \cdot [\varphi_{i_1}(p) \wedge \cdots \wedge \varphi_{i_k}(p)]$$

for certain functions $\omega_{i_1,\ldots,i_k}$; the form ω is called continuous, differentiable, etc., if these functions are. We shall usually assume tacitly that forms and vector fields are differentiable, and "differentiable" will henceforth mean "C^∞"; this is a simplifying assumption that eliminates the need for counting how many times a function is differentiated in a proof. The sum $\omega + \eta$, product $f \cdot \omega$, and wedge product $\omega \wedge \eta$ are defined in the obvious way. A function f is considered to be a 0-form and $f \cdot \omega$ is also written $f \wedge \omega$.

If $f\colon \mathbf{R}^n \to \mathbf{R}$ is differentiable, then $Df(p) \in \Lambda^1(\mathbf{R}^n)$. By a minor modification we therefore obtain a 1-form df, defined by

$$df(p)(v_p) = Df(p)(v).$$

Let us consider in particular the 1-forms $d\pi^i$. It is customary to let x^i denote the *function* π^i. (On $\mathbf{R}^3$ we often denote x^1, x^2, and x^3 by x, y, and z.) This standard notation has obvious disadvantages but it allows many classical results to be expressed by formulas of equally classical appearance. Since $dx^i(p)(v_p) = d\pi^i(p)(v_p) = D\pi^i(p)(v) = v^i$, we see that $dx^1(p), \dots, dx^n(p)$ is just the dual basis to $(e_1)_p, \dots, (e_n)_p$. Thus every k-form ω can be written

$$\omega = \sum_{i_1 < \cdots < i_k} \omega_{i_1, \dots, i_k} dx^{i_1} \wedge \cdots \wedge dx^{i_k}.$$

The expression for df is of particular interest.

4-7 Theorem. *If $f\colon \mathbf{R}^n \to \mathbf{R}$ is differentiable, then*

$$df = D_1 f \cdot dx^1 + \cdots + D_n f \cdot dx^n.$$

In classical notation,

$$df = \frac{\partial f}{\partial x^1} dx^1 + \cdots + \frac{\partial f}{\partial x^n} dx^n.$$

Proof. $df(p)(v_p) = Df(p)(v) = \Sigma_{i=1}^n v^i \cdot D_i f(p)$
$= \Sigma_{i=1}^n dx^i(p)(v_p) \cdot D_i f(p)$. ▌

If we consider now a differentiable function $f\colon \mathbf{R}^n \to \mathbf{R}^m$ we have a linear transformation $Df(p)\colon \mathbf{R}^n \to \mathbf{R}^m$. Another minor modification therefore produces a linear transformation $f_*\colon \mathbf{R}^n{}_p \to \mathbf{R}^m{}_{f(p)}$ defined by

$$f_*(v_p) = (Df(p)(v))_{f(p)}.$$

This linear transformation induces a linear transformation $f^*\colon \Lambda^k(\mathbf{R}^m{}_{f(p)}) \to \Lambda^k(\mathbf{R}^n{}_p)$. If ω is a k-form on $\mathbf{R}^m$ we can therefore define a k-form $f^*\omega$ on $\mathbf{R}^n$ by $(f^*\omega)(p) = f^*(\omega(f(p)))$.

Recall this means that if $v_1, \ldots, v_k \in \mathbf{R}^n{}_p$, then we have $f^*\omega(p)(v_1, \ldots, v_k) = \omega(f(p))(f_*(v_1), \ldots, f_*(v_k))$. As an antidote to the abstractness of these definitions we present a theorem, summarizing the important properties of f^*, which allows explicit calculations of $f^*\omega$.

4-8 Theorem. *If $f\colon \mathbf{R}^n \to \mathbf{R}^m$ is differentiable, then*

(1) $f^*(dx^i) = \sum_{j=1}^n D_jf^i \cdot dx^j = \sum_{j=1}^n \dfrac{\partial f^i}{\partial x^j}\, dx^j$.

(2) $f^*(\omega_1 + \omega_2) = f^*(\omega_1) + f^*(\omega_2)$.

(3) $f^*(g \cdot \omega) = (g \circ f) \cdot f^*\omega$.

(4) $f^*(\omega \wedge \eta) = f^*\omega \wedge f^*\eta$.

Proof

(1)
$$\begin{aligned} f^*(dx^i)(p)(v_p) &= dx^i(f(p))(f_*v_p) \\ &= dx^i(f(p))\Big(\textstyle\sum_{j=1}^n v^j \cdot D_jf^1(p), \ldots, \sum_{j=1}^n v^j \cdot D_jf^m(p)\Big)_{f(p)} \\ &= \textstyle\sum_{j=1}^n v^j \cdot D_jf^i(p) \\ &= \textstyle\sum_{j=1}^n D_jf^i(p) \cdot dx^j(p)(v_p). \end{aligned}$$

The proofs of (2), (3), and (4) are left to the reader. ▌

By repeatedly applying Theorem 4-8 we have, for example,

$$\begin{aligned} f^*(P\,dx^1 \wedge dx^2 + Q\,dx^2 \wedge dx^3) = (P \circ f)[f^*(dx^1) \wedge f^*(dx^2)] \\ + (Q \circ f)[f^*(dx^2) \wedge f^*(dx^3)]. \end{aligned}$$

The expression obtained by expanding out each $f^*(dx^i)$ is quite complicated. (It is helpful to remember, however, that we have $dx^i \wedge dx^i = (-1)dx^i \wedge dx^i = 0$.) In one special case it will be worth our while to make an explicit evaluation.

4-9 Theorem. *If $f\colon \mathbf{R}^n \to \mathbf{R}^n$ is differentiable, then*

$$f^*(h\,dx^1 \wedge \cdots \wedge dx^n) = (h \circ f)(\det f')\,dx^1 \wedge \cdots \wedge dx^n.$$

Proof. Since

$$f^*(h\,dx^1 \wedge \cdots \wedge dx^n) = (h \circ f)f^*(dx^1 \wedge \cdots \wedge dx^n),$$

it suffices to show that

$$f^*(dx^1 \wedge \cdots \wedge dx^n) = (\det f')\, dx^1 \wedge \cdots \wedge dx^n.$$

Let $p \in \mathbf{R}^n$ and let $A = (a_{ij})$ be the matrix of $f'(p)$. Here, and whenever convenient and not confusing, we shall omit "p" in $dx^1 \wedge \cdots \wedge dx^n(p)$, etc. Then

$$\begin{aligned} f^*(dx^1 \wedge \cdots \wedge dx^n)(e_1, \ldots, e_n) &\\ &= dx^1 \wedge \cdots \wedge dx^n(f_*e_1, \ldots, f_*e_n) \\ &= dx^1 \wedge \cdots \wedge \left(\sum_{i=1}^n a_{i1}e_i, \ldots, \sum_{i=1}^n a_{in}e_i\right) \\ &= \det(a_{ij}) \cdot dx^1 \wedge \cdots \wedge dx^n(e_1, \ldots, e_n), \end{aligned}$$

by Theorem 4-6. ∎

An important construction associated with forms is a generalization of the operator d which changes 0-forms into 1-forms. If

$$\omega = \sum_{i_1 < \cdots < i_k} \omega_{i_1, \ldots, i_k}\, dx^{i_1} \wedge \cdots \wedge dx^{i_k},$$

we define a $(k+1)$-form $d\omega$, the **differential** of ω, by

$$\begin{aligned} d\omega &= \sum_{i_1 < \cdots < i_k} d\omega_{i_1, \ldots, i_k} \wedge dx^{i_1} \wedge \cdots \wedge dx^{i_k} \\ &= \sum_{i_1 < \cdots < i_k} \sum_{\alpha=1}^n D_\alpha(\omega_{i_1, \ldots, i_k}) \cdot dx^\alpha \wedge dx^{i_1} \wedge \cdots \wedge dx^{i_k}. \end{aligned}$$

4-10 Theorem

(1) $d(\omega + \eta) = d\omega + d\eta$.
(2) *If ω is a k-form and η is an l-form, then*

$$d(\omega \wedge \eta) = d\omega \wedge \eta + (-1)^k \omega \wedge d\eta.$$

(3) $d(d\omega) = 0$. *Briefly,* $d^2 = 0$.
(4) *If ω is a k-form on $\mathbf{R}^m$ and $f\colon \mathbf{R}^n \to \mathbf{R}^m$ is differentiable, then $f^*(d\omega) = d(f^*\omega)$.*

Proof

(1) Left to the reader.

(2) The formula is true if $\omega = dx^{i_1} \wedge \cdots \wedge dx^{i_k}$ and $\eta = dx^{j_1} \wedge \cdots \wedge dx^{j_l}$, since all terms vanish. The formula is easily checked when ω is a 0-form. The general formula may be derived from (1) and these two observations.

(3) Since

$$d\omega = \sum_{i_1 < \cdots < i_k} \sum_{\alpha=1}^{n} D_\alpha(\omega_{i_1, \ldots, i_k}) dx^\alpha \wedge dx^{i_1} \wedge \cdots \wedge dx^{i_k},$$

we have

$$d(d\omega) = \sum_{i_1 < \cdots < i_k} \sum_{\alpha=1}^{n} \sum_{\beta=1}^{n} D_{\alpha,\beta}(\omega_{i_1, \ldots, i_k}) dx^\beta \wedge dx^\alpha \wedge dx^{i_1} \wedge \cdots \wedge dx^{i_k}.$$

In this sum the terms

$$D_{\alpha,\beta}(\omega_{i_1, \ldots, i_k}) dx^\beta \wedge dx^\alpha \wedge dx^{i_1} \wedge \cdots \wedge dx^{i_k}$$

and

$$D_{\beta,\alpha}(\omega_{i_1, \ldots, i_k}) dx^\alpha \wedge dx^\beta \wedge dx^{i_1} \wedge \cdots \wedge dx^{i_k}$$

cancel in pairs.

(4) This is clear if ω is a 0-form. Suppose, inductively, that (4) is true when ω is a k-form. It suffices to prove (4) for a $(k+1)$-form of the type $\omega \wedge dx^i$. We have

$$\begin{aligned} f^*(d(\omega \wedge dx^i)) &= f^*(d\omega \wedge dx^i + (-1)^k \omega \wedge d(dx^i)) \\ &= f^*(d\omega \wedge dx^i) = f^*(d\omega) \wedge f^*(dx^i) \\ &= d(f^*\omega \wedge f^*(dx^i)) \qquad \text{by (2) and (3)} \\ &= d(f^*(\omega \wedge dx^i)). \quad \blacksquare \end{aligned}$$

A form ω is called **closed** if $d\omega = 0$ and **exact** if $\omega = d\eta$, for some η. Theorem 4-10 shows that every exact form is closed, and it is natural to ask whether, conversely, every closed form is exact. If ω is the 1-form $P\,dx + Q\,dy$ on $\mathbf{R}^2$, then

$$\begin{aligned} d\omega &= (D_1P\,dx + D_2P\,dy) \wedge dx + (D_1Q\,dx + D_2Q\,dy) \wedge dy \\ &= (D_1Q - D_2P) dx \wedge dy. \end{aligned}$$

Thus, if $d\omega = 0$, then $D_1Q = D_2P$. Problems 2-21 and 3-34 show that there is a 0-form f such that $\omega = df = D_1f\,dx + D_2f\,dy$. If ω is defined only on a subset of $\mathbf{R}^2$, however, such a function may not exist. The classical example is the form

$$\omega = \frac{-y}{x^2 + y^2}\,dx + \frac{x}{x^2 + y^2}\,dy$$

defined on $\mathbf{R}^2 - 0$. This form is usually denoted $d\theta$ (where θ is defined in Problem 3-41), since (Problem 4-21) it equals $d\theta$ on the set $\{(x,y)\colon x < 0, \text{ or } x \geq 0 \text{ and } y \neq 0\}$, where θ is defined. Note, however, that θ cannot be defined continuously on all of $\mathbf{R}^2 - 0$. If $\omega = df$ for some function $f\colon \mathbf{R}^2 - 0 \to \mathbf{R}$, then $D_1f = D_1\theta$ and $D_2f = D_2\theta$, so $f = \theta + \text{constant}$, showing that such an f cannot exist.

Suppose that $\omega = \sum_{i=1}^n \omega_i\,dx^i$ is a 1-form on $\mathbf{R}^n$ and ω happens to equal $df = \sum_{i=1}^n D_if \cdot dx^i$. We can clearly assume that $f(0) = 0$. As in Problem 2-35, we have

$$\begin{aligned} f(x) &= \int_0^1 \frac{d}{dt} f(tx)\,dt \\ &= \int_0^1 \sum_{i=1}^n D_if(tx) \cdot x^i\,dt \\ &= \int_0^1 \sum_{i=1}^n \omega_i(tx) \cdot x^i\,dt. \end{aligned}$$

This suggests that in order to find f, given ω, we consider the function $I\omega$, defined by

$$I\omega(x) = \int_0^1 \sum_{i=1}^n \omega_i(tx) \cdot x^i\,dt.$$

Note that the definition of $I\omega$ makes sense if ω is defined only on an open set $A \subset \mathbf{R}^n$ with the property that whenever $x \in A$, the line segment from 0 to x is contained in A; such an open set is called **star-shaped** with respect to 0 (Figure 4-3). A somewhat involved calculation shows that (on a star-shaped open set) we have $\omega = d(I\omega)$ provided that ω satisfies the necessary condition $d\omega = 0$. The calculation, as well as the definition of $I\omega$, may be generalized considerably:

FIGURE 4-3

4-11 Theorem (Poincaré Lemma). *If $A \subset \mathbf{R}^n$ is an open set star-shaped with respect to 0, then every closed form on A is exact.*

Proof. We will define a function I from l-forms to $(l-1)$-forms (for each l), such that $I(0) = 0$ and $\omega = I(d\omega) + d(I\omega)$ for any form ω. It follows that $\omega = d(I\omega)$ if $d\omega = 0$. Let

$$\omega = \sum_{i_1 < \cdots < i_l} \omega_{i_1, \ldots, i_l}\, dx^{i_1} \wedge \cdots \wedge dx^{i_l}.$$

Since A is star-shaped we can define

$$I\omega(x) = \sum_{i_1 < \cdots < i_l} \sum_{\alpha=1}^{l} (-1)^{\alpha-1} \left(\int_0^1 t^{l-1} \omega_{i_1, \ldots, i_l}(tx)dt \right) x^{i_\alpha} \\ dx^{i_1} \wedge \cdots \wedge \widehat{dx^{i_\alpha}} \wedge \cdots \wedge dx^{i_l}.$$

(The symbol $\frown$ over dx^{i_α} indicates that it is omitted.) The proof that $\omega = I(d\omega) + d(I\omega)$ is an elaborate computation: We have, using Problem 3-32,

$$d(I\omega) = l \cdot \sum_{i_1 < \cdots < i_l} \left(\int_0^1 t^{l-1} \omega_{i_1, \ldots, i_l}(tx)dt \right) dx^{i_1} \wedge \cdots \wedge dx^{i_l}$$

$$+ \sum_{i_1 < \cdots < i_l} \sum_{\alpha=1}^{l} \sum_{j=1}^{n} (-1)^{\alpha-1} \left(\int_0^1 t^l D_j(\omega_{i_1, \ldots, i_l})(tx)dt \right) x^{i_\alpha} dx^j \wedge dx^{i_1} \wedge \cdots \wedge \widehat{dx^{i_\alpha}} \wedge \cdots \wedge dx^{i_l}.$$

(Explain why we have the factor t^l, instead of t^{l-1}.) We also have

$$d\omega = \sum_{i_1 < \cdots < i_l} \sum_{j=1}^{n} D_j(\omega_{i_1, \ldots, i_l}) \cdot dx^j \wedge dx^{i_1} \wedge \cdots \wedge dx^{i_l}.$$

Applying I to the $(l+1)$-form $d\omega$, we obtain

$$I(d\omega) = \sum_{i_1 < \cdots < i_l} \sum_{j=1}^{n} \left(\int_0^1 t^l D_j(\omega_{i_1, \ldots, i_l})(tx)dt \right) x^j dx^{i_1} \wedge \cdots \wedge dx^{i_l}$$

$$- \sum_{i_1 < \cdots < i_l} \sum_{j=1}^{n} \sum_{\alpha=1}^{l} (-1)^{\alpha-1} \left(\int_0^1 t^l D_j(\omega_{i_1, \ldots, i_l})(tx)dt \right) x^{i_\alpha} dx^j \wedge dx^{i_1} \wedge \cdots \wedge \widehat{dx^{i_\alpha}} \wedge \cdots \wedge dx^{i_l}.$$

Adding, the triple sums cancel, and we obtain

$$\begin{aligned} d(I\omega) + I(d\omega) &= \sum_{i_1 < \cdots < i_l} l \cdot \left(\int_0^1 t^{l-1} \omega_{i_1, \ldots, i_l}(tx)dt \right) dx^{i_1} \wedge \cdots \wedge dx^{i_l} \\ &\quad + \sum_{i_1 < \cdots < i_l} \sum_{j=1}^{n} \left(\int_0^1 t^l x^j D_j(\omega_{i_1, \ldots, i_l})(tx)dt \right) dx^{i_1} \wedge \cdots \wedge dx^{i_l} \\ &= \sum_{i_1 < \cdots < i_l} \left(\int_0^1 \frac{d}{dt} [t^l \omega_{i_1, \ldots, i_l}(tx)]dt \right) dx^{i_1} \wedge \cdots \wedge dx^{i_l} \\ &= \sum_{i_1 < \cdots < i_l} \omega_{i_1, \ldots, i_l} dx^{i_1} \wedge \cdots \wedge dx^{i_l} \\ &= \omega. \quad \blacksquare \end{aligned}$$

Problems. **4-13.** (a) If $f\colon \mathbf{R}^n \to \mathbf{R}^m$ and $g\colon \mathbf{R}^m \to \mathbf{R}^p$, show that $(g \circ f)_* = g_* \circ f_*$ and $(g \circ f)^* = f^* \circ g^*$.

(b) If $f,g\colon \mathbf{R}^n \to \mathbf{R}$, show that $d(f \cdot g) = f \cdot dg + g \cdot df$.

4-14. Let c be a differentiable curve in $\mathbf{R}^n$, that is, a differentiable function $c\colon [0,1] \to \mathbf{R}^n$. Define the **tangent vector** v of c at t as $c_*((e_1)_t) = ((c^1)'(t), \ldots, (c^n)'(t))_{c(t)}$. If $f\colon \mathbf{R}^n \to \mathbf{R}^m$, show that the tangent vector to $f \circ c$ at t is $f_*(v)$.

4-15. Let $f\colon \mathbf{R} \to \mathbf{R}$ and define $c\colon \mathbf{R} \to \mathbf{R}^2$ by $c(t) = (t,f(t))$. Show that the end point of the tangent vector of c at t lies on the tangent line to the graph of f at $(t,f(t))$.

4-16. Let $c\colon [0,1] \to \mathbf{R}^n$ be a curve such that $|c(t)| = 1$ for all t. Show that $c(t)_{c(t)}$ and the tangent vector to c at t are perpendicular.

4-17. If $f\colon \mathbf{R}^n \to \mathbf{R}^n$, define a vector field $\mathbf{f}$ by $\mathbf{f}(p) = f(p)_p \in \mathbf{R}^n{}_p$.

(a) Show that every vector field F on $\mathbf{R}^n$ is of the form $\mathbf{f}$ for some f.

(b) Show that $\operatorname{div} \mathbf{f} = \operatorname{trace} f'$.

4-18. If $f\colon \mathbf{R}^n \to \mathbf{R}$, define a vector field $\operatorname{grad} f$ by

$$(\operatorname{grad} f)(p) = D_1 f(p) \cdot (e_1)_p + \cdots + D_n f(p) \cdot (e_n)_p.$$

For obvious reasons we also write $\operatorname{grad} f = \nabla f$. If $\nabla f(p) = w_p$, prove that $D_v f(p) = \langle v,w \rangle$ and conclude that $\nabla f(p)$ is the direction in which f is changing fastest at p.

4-19. If F is a vector field on $\mathbf{R}^3$, define the forms

$$\omega_F^1 = F^1\, dx + F^2\, dy + F^3\, dz,$$
$$\omega_F^2 = F^1\, dy \wedge dz + F^2\, dz \wedge dx + F^3\, dx \wedge dy.$$

(a) Prove that

$$df = \omega^1_{\operatorname{grad} f},$$
$$d(\omega_F^1) = \omega^2_{\operatorname{curl} F},$$
$$d(\omega_F^2) = (\operatorname{div} F)\, dx \wedge dy \wedge dz.$$

(b) Use (a) to prove that

$$\operatorname{curl} \operatorname{grad} f = 0,$$
$$\operatorname{div} \operatorname{curl} F = 0.$$

(c) If F is a vector field on a star-shaped open set A and $\operatorname{curl} F = 0$, show that $F = \operatorname{grad} f$ for some function $f\colon A \to \mathbf{R}$. Similarly, if $\operatorname{div} F = 0$, show that $F = \operatorname{curl} G$ for some vector field G on A.

4-20. Let $f\colon U \to \mathbf{R}^n$ be a differentiable function with a differentiable inverse $f^{-1}\colon f(U) \to \mathbf{R}^n$. If every closed form on U is exact, show that the same is true for $f(U)$. *Hint:* If $d\omega = 0$ and $f^*\omega = d\eta$, consider $(f^{-1})^*\eta$.

4-21.* Prove that on the set where θ is defined we have

$$d\theta = \frac{-y}{x^2 + y^2}\,dx + \frac{x}{x^2 + y^2}\,dy.$$

GEOMETRIC PRELIMINARIES

A **singular n-cube** in $A \subset \mathbf{R}^n$ is a continuous function c: $[0,1]^n \to A$ (here $[0,1]^n$ denotes the n-fold product $[0,1] \times \cdots \times [0,1]$). We let $\mathbf{R}^0$ and $[0,1]^0$ both denote $\{0\}$. A singular 0-cube in A is then a function f: $\{0\} \to A$ or, what amounts to the same thing, a point in A. A singular 1-cube is often called a **curve.** A particularly simple, but particularly important example of a singular n-cube in $\mathbf{R}^n$ is the **standard n-cube** I^n: $[0,1]^n \to \mathbf{R}^n$ defined by $I^n(x) = x$ for $x \in [0,1]^n$.

We shall need to consider formal sums of singular n-cubes in A multiplied by integers, that is, expressions like

$$2c_1 + 3c_2 - 4c_3,$$

where c_1, c_2, c_3 are singular n-cubes in A. Such a finite sum of singular n-cubes with integer coefficients is called an **n-chain** in A. In particular a singular n-cube c is also considered as an n-chain $1 \cdot c$. It is clear how n-chains can be added, and multiplied by integers. For example

$$2(c_1 + 3c_4) + (-2)(c_1 + c_3 + c_2) = -2c_2 - 2c_3 + 6c_4.$$

(A rigorous exposition of this formalism is presented in Problem 4-22.)

For each singular n-chain c in A we shall define an $(n-1)$-chain in A called the **boundary** of c and denoted ∂c. The boundary of I^2, for example, might be defined as the sum of four singular 1-cubes arranged counterclockwise around the boundary of $[0,1]^2$, as indicated in Figure 4-4(a). It is actually much more convenient to define ∂I^2 as the sum, with the indicated coefficients, of the four singular 1-cubes shown in Figure 4-4(b). The precise definition of ∂I^n requires some preliminary notions. For each i with $1 \le i \le n$ we define two singular $(n-1)$-cubes $I^n_{(i,0)}$ and $I^n_{(i,1)}$ as follows. If

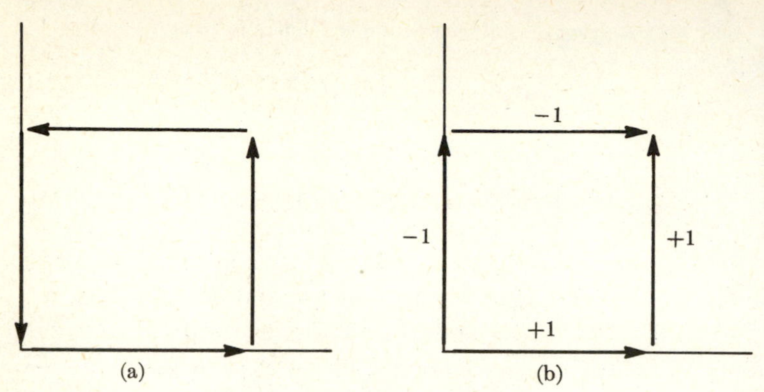

FIGURE 4-4

$x \in [0,1]^{n-1}$, then

$$\begin{aligned} I^n_{(i,0)}(x) &= I^n(x^1, \ldots, x^{i-1}, 0, x^i, \ldots, x^{n-1}) \\ &= (x^1, \ldots, x^{i-1}, 0, x^i, \ldots, x^{n-1}), \\ I^n_{(i,1)}(x) &= I^n(x^1, \ldots, x^{i-1}, 1, x^i, \ldots, x^{n-1}) \\ &= (x^1, \ldots, x^{i-1}, 1, x^i, \ldots, x^{n-1}). \end{aligned}$$

We call $I^n_{(i,0)}$ the $(i,0)$-face of I^n and $I^n_{(i,1)}$ the $(i,1)$-face (Figure 4-5). We then define

$$\partial I^n = \sum_{i=1}^{n} \sum_{\alpha=0,1} (-1)^{i+\alpha} I^n_{(i,\alpha)}.$$

For a general singular n-cube $c\colon [0,1]^n \to A$ we first define the (i,α)-face,

$$c_{(i,\alpha)} = c \circ (I^n_{(i,\alpha)})$$

and then define

$$\partial c = \sum_{i=1}^{n} \sum_{\alpha=0,1} (-1)^{i+\alpha} c_{(i,\alpha)}.$$

Finally we define the boundary of an n-chain $\Sigma a_i c_i$ by

$$\partial(\Sigma a_i c_i) = \Sigma a_i \partial(c_i).$$

Although these few definitions suffice for all applications in this book, we include here the one standard property of ∂.

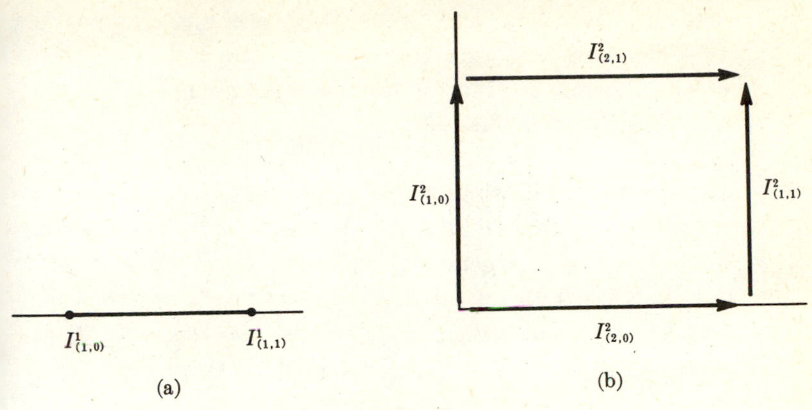

FIGURE 4-5

4-12 Theorem. *If c is an n-chain in A, then $\partial(\partial c) = 0$. Briefly, $\partial^2 = 0$.*

Proof. Let $i \leq j$ and consider $(I^n_{(i,\alpha)})_{(j,\beta)}$. If $x \in [0,1]^{n-2}$, then, remembering the definition of the (j,β)-face of a singular n-cube, we have

$$\begin{aligned}(I^n_{(i,\alpha)})_{(j,\beta)}(x) &= I^n_{(i,\alpha)}(I^{n-1}_{(j,\beta)}(x)) \\ &= I^n_{(i,\alpha)}(x^1, \ldots, x^{j-1}, \beta, x^j, \ldots, x^{n-2}) \\ &= I^n(x^1, \ldots, x^{i-1}, \alpha, x^i, \ldots, x^{j-1}, \beta, x^j, \ldots, x^{n-2}).\end{aligned}$$

Similarly

$$\begin{aligned}(I^n_{(j+1,\beta)})_{(i,\alpha)} &= I^n_{(j+1,\beta)}(I^{n-1}_{(i,\alpha)}(x)) \\ &= I^n_{(j+1,\beta)}(x^1, \ldots, x^{i-1}, \alpha, x^i, \ldots, x^{n-2}) \\ &= I^n(x^1, \ldots, x^{i-1}, \alpha, x^i, \ldots, x^{j-1}, \beta, x^j, \ldots, x^{n-2}).\end{aligned}$$

Thus $(I^n_{(i,\alpha)})_{(j,\beta)} = (I^n_{(j+1,\beta)})_{(i,\alpha)}$ for $i \leq j$. (It may help to verify this in Figure 4-5.) It follows easily for any singular n-cube c that $(c_{(i,\alpha)})_{(j,\beta)} = (c_{(j+1,\beta)})_{(i,\alpha)}$ when $i \leq j$. Now

$$\begin{aligned}\partial(\partial c) &= \partial\left(\sum_{i=1}^{n} \sum_{\alpha=0,1} (-1)^{i+\alpha} c_{(i,\alpha)}\right) \\ &= \sum_{i=1}^{n} \sum_{\alpha=0,1} \sum_{j=1}^{n-1} \sum_{\beta=0,1} (-1)^{i+\alpha+j+\beta} (c_{(i,\alpha)})_{(j,\beta)}.\end{aligned}$$

In this sum $(c_{(i,\alpha)})_{(j,\beta)}$ and $(c_{(j+1,\beta)})_{(i,\alpha)}$ occur with opposite signs. Therefore all terms cancel out in pairs and $\partial(\partial c) = 0$. Since the theorem is true for any singular n-cube, it is also true for singular n-chains. ▌

It is natural to ask whether Theorem 4-12 has a converse: If $\partial c = 0$, is there a chain d in A such that $c = \partial d$? The answer depends on A and is generally "no." For example, define $c\colon [0,1] \to \mathbf{R}^2 - 0$ by $c(t) = (\sin 2\pi nt, \cos 2\pi nt)$, where n is a non-zero integer. Then $c(1) = c(0)$, so $\partial c = 0$. But (Problem 4-26) there is no 2-chain c' in $\mathbf{R}^2 - 0$, with $\partial c' = c$.

Problems. **4-22.** Let $\mathcal{S}$ be the set of all singular n-cubes, and $\mathbf{Z}$ the integers. An ***n*-chain** is a function $f\colon \mathcal{S} \to \mathbf{Z}$ such that $f(c) = 0$ for all but finitely many c. Define $f + g$ and nf by $(f + g)(c) = f(c) + g(c)$ and $nf(c) = n \cdot f(c)$. Show that $f + g$ and nf are n-chains if f and g are. If $c \in \mathcal{S}$, let c also denote the function f such that $f(c) = 1$ and $f(c') = 0$ for $c' \neq c$. Show that every n-chain f can be written $a_1c_1 + \cdots + a_kc_k$ for some integers $a_1, \ldots, a_k$ and singular n-cubes $c_1, \ldots, c_k$.

4-23. For $R > 0$ and n an integer, define the singular 1-cube $c_{R,n}\colon [0,1] \to \mathbf{R}^2 - 0$ by $c_{R,n}(t) = (R \cos 2\pi nt, R \sin 2\pi nt)$. Show that there is a singular 2-cube $c\colon [0,1]^2 \to \mathbf{R}^2 - 0$ such that $c_{R_1,n} - c_{R_2,n} = \partial c$.

4-24. If c is a singular 1-cube in $\mathbf{R}^2 - 0$ with $c(0) = c(1)$, show that there is an integer n such that $c - c_{1,n} = \partial c^2$ for some 2-chain c^2. *Hint:* First partition $[0,1]$ so that each $c([t_{i-1},t_i])$ is contained on one side of some line through 0.

THE FUNDAMENTAL THEOREM OF CALCULUS

The fact that $d^2 = 0$ and $\partial^2 = 0$, not to mention the typographical similarity of d and ∂, suggests some connection between chains and forms. This connection is established by integrating forms over chains. Henceforth only differentiable singular n-cubes will be considered.

If ω is a k-form on $[0,1]^k$, then $\omega = f\,dx^1 \wedge \cdots \wedge dx^k$ for a unique function f. We define

$$\int_{[0,1]^k} \omega = \int_{[0,1]^k} f.$$

We could also write this as

$$\int_{[0,1]^k} f\,dx^1 \wedge \cdots \wedge dx^k = \int_{[0,1]^k} f(x^1, \ldots, x^k)dx^1 \cdots dx^k,$$

one of the reasons for introducing the functions x^i.

If ω is a k-form on A and c is a singular k-cube in A, we define

$$\int_c \omega = \int_{[0,1]^k} c^*\omega.$$

Note, in particular, that

$$\begin{aligned}\int_{I^k} f\,dx^1 \wedge \cdots \wedge dx^k &= \int_{[0,1]^k} (I^k)^*(f\,dx^1 \wedge \cdots \wedge dx^k) \\ &= \int_{[0,1]^k} f(x^1, \ldots, x^k)dx^1 \cdots dx^k.\end{aligned}$$

A special definition must be made for $k = 0$. A 0-form ω is a function; if $c\colon \{0\} \to A$ is a singular 0-cube in A we define

$$\int_c \omega = \omega(c(0)).$$

The integral of ω over a k-chain $c = \Sigma a_i c_i$ is defined by

$$\int_c \omega = \sum a_i \int_{c_i} \omega.$$

The integral of a 1-form over a 1-chain is often called a **line integral.** If $P\,dx + Q\,dy$ is a 1-form on $\mathbf{R}^2$ and $c\colon [0,1] \to \mathbf{R}^2$ is a singular 1-cube (a curve), then one can (but we will not) prove that

$$\begin{aligned}\int_c P\,dx + Q\,dy = \lim \sum_{i=1}^{n} &[c^1(t_i) - c^1(t_{i-1})] \cdot P(c(t^i)) \\ &+ [c^2(t_i) - c^2(t_{i-1})] \cdot Q(c(t^i))\end{aligned}$$

where $t_0, \ldots, t_n$ is a partition of $[0,1]$, the choice of t^i in $[t_{i-1}, t_i]$ is arbitrary, and the limit is taken over all partitions as the maximum of $|t_i - t_{i-1}|$ goes to 0. The right side is often taken as a definition of $\int_c P\,dx + Q\,dy$. This is a natural definition to make, since these sums are very much like the sums appearing in the definition of ordinary integrals. However such an expression is almost impossible to work with and is quickly equated with an integral equivalent to $\int_{[0,1]} c^*(P\,dx + Q\,dy)$. Analogous definitions for **surface integrals,** that is, integrals of 2-forms over singular 2-cubes, are even more complicated and difficult to use. This is one reason why we have avoided such an approach. The other reason is that the definition given here is the one that makes sense in the more general situations considered in Chapter 5.

The relationship between forms, chains, d, and ∂ is summed up in the neatest possible way by Stokes' theorem, sometimes called the fundamental theorem of calculus in higher dimensions (if $k = 1$ and $c = I^1$, it really is the fundamental theorem of calculus).

4-13 Theorem (Stokes' Theorem). *If ω is a $(k-1)$-form on an open set $A \subset \mathbf{R}^n$ and c is a k-chain in A, then*

$$\int_c d\omega = \int_{\partial c} \omega.$$

Proof. Suppose first that $c = I^k$ and ω is a $(k-1)$-form on $[0,1]^k$. Then ω is the sum of $(k-1)$-forms of the type

$$f\,dx^1 \wedge \cdots \wedge \widehat{dx^i} \wedge \cdots \wedge dx^k,$$

and it suffices to prove the theorem for each of these. This simply involves a computation:

Note that

$$\int_{[0,1]^{k-1}} I^k_{(j,\alpha)}{}^*(f\,dx^1 \wedge \cdots \wedge \widehat{dx^i} \wedge \cdots \wedge dx^k)$$
$$= \begin{cases} 0 & \text{if } j \neq i, \\ \displaystyle\int_{[0,1]^k} f(x^1, \ldots, \alpha, \ldots, x^k)dx^1 \cdots dx^k & \text{if } j = i. \end{cases}$$

Therefore

$$\int_{\partial I^k} f\,dx^1 \wedge \cdots \wedge \widehat{dx^i} \wedge \cdots \wedge dx^k$$

$$= \sum_{j=1}^{k} \sum_{\alpha=0,1} (-1)^{j+\alpha} \int_{[0,1]^{k-1}} I^k_{(j,\alpha)}{}^*(f\,dx^1 \wedge \cdots \wedge \widehat{dx^i} \wedge \cdots \wedge dx^k)$$

$$= (-1)^{i+1} \int_{[0,1]^k} f(x^1, \ldots, 1, \ldots, x^k)dx^1 \cdots dx^k + (-1)^i \int_{[0,1]^k} f(x^1, \ldots, 0, \ldots, x^k)dx^1 \cdots dx^k.$$

On the other hand,

$$\int_{I^k} d(f\,dx^1 \wedge \cdots \wedge \widehat{dx^i} \wedge \cdots \wedge dx^k) = \int_{[0,1]^k} D_i f\,dx^i \wedge dx^1 \wedge \cdots \wedge \widehat{dx^i} \wedge \cdots \wedge dx^k = (-1)^{i-1} \int_{[0,1]^k} D_i f.$$

By Fubini's theorem and the fundamental theorem of calculus (in one dimension) we have

$$\int_{I^k} d(f\,dx^1 \wedge \cdots \wedge \widehat{dx^i} \wedge \cdots \wedge dx^k)$$

$$= (-1)^{i-1} \int_0^1 \cdots \left(\int_0^1 D_i f(x^1, \ldots, x^k)dx^i\right) dx^1 \cdots \widehat{dx^i} \cdots dx^k$$

$$= (-1)^{i-1} \int_0^1 \cdots \int_0^1 [f(x^1, \ldots, 1, \ldots, x^k) - f(x^1, \ldots, 0, \ldots, x^k)]dx^1 \cdots \widehat{dx^i} \cdots dx^k$$

$$= (-1)^{i-1} \int_{[0,1]^k} f(x^1, \ldots, 1, \ldots, x^k)dx^1 \cdots dx^k + (-1)^i \int_{[0,1]^k} f(x^1, \ldots, 0, \ldots, x^k)dx^1 \cdots dx^k.$$

Thus

$$\int_{I^k} d\omega = \int_{\partial I^k} \omega.$$

If c is an arbitrary singular k-cube, working through the definitions will show that

$$\int_{\partial c} \omega = \int_{\partial I^k} c^*\omega.$$

Therefore

$$\int_c d\omega = \int_{I^k} c^*(d\omega) = \int_{I^k} d(c^*\omega) = \int_{\partial I^k} c^*\omega = \int_{\partial c} \omega.$$

Finally, if c is a k-chain $\Sigma a_i c_i$, we have

$$\int_c d\omega = \sum a_i \int_{c_i} d\omega = \sum a_i \int_{\partial c_i} \omega = \int_{\partial c} \omega. \blacksquare$$

Stokes' theorem shares three important attributes with many fully evolved major theorems:

1. It is trivial.
2. It is trivial because the terms appearing in it have been properly defined.
3. It has significant consequences.

Since this entire chapter was little more than a series of definitions which made the statement and proof of Stokes' theorem possible, the reader should be willing to grant the first two of these attributes to Stokes' theorem. The rest of the book is devoted to justifying the third.

Problems. **4-25.** (*Independence of parameterization*). Let c be a singular k-cube and p: $[0,1]^k \to [0,1]^k$ a 1-1 function such that $p([0,1]^k) = [0,1]^k$ and $\det p'(x) \geq 0$ for $x \in [0,1]^k$. If ω is a k-form, show that

$$\int_c \omega = \int_{c \circ p} \omega.$$

4-26. Show that $\int_{c_{R,n}} d\theta = 2\pi n$, and use Stokes' theorem to conclude that $c_{R,n} \neq \partial c$ for any 2-chain c in $\mathbf{R}^2 - 0$ (recall the definition of $c_{R,n}$ in Problem 4-23).

4-27. Show that the integer n of Problem 4-24 is unique. This integer is called the **winding number** of c around 0.

4-28. Recall that the set of complex numbers $\mathbf{C}$ is simply $\mathbf{R}^2$ with $(a,b) = a + bi$. If $a_1, \ldots, a_n \in \mathbf{C}$ let f: $\mathbf{C} \to \mathbf{C}$ be $f(z) = z^n + a_1 z^{n-1} + \cdots + a_n$. Define the singular 1-cube $c_{R,f}$:

$[0,1] \to \mathbf{C} - 0$ by $c_{R,f} = f \circ c_{R,1}$, and the singular 2-cube c by $c(s,t) = t \cdot c_{R,n}(s) + (1 - t)c_{R,f}(s)$.

(a) Show that $\partial c = c_{R,f} - c_{R,n}$, and that $c([0,1] \times [0,1]) \subset \mathbf{C} - 0$ if R is large enough.

(b) Using Problem 4-26, prove the *Fundamental Theorem of Algebra:* Every polynomial $z^n + a_1 z^{n-1} + \cdots + a_n$ with $a_i \in \mathbf{C}$ has a root in $\mathbf{C}$.

4-29. If ω is a 1-form $f\,dx$ on $[0,1]$ with $f(0) = f(1)$, show that there is a unique number λ such that $\omega - \lambda\,dx = dg$ for some function g with $g(0) = g(1)$. *Hint:* Integrate $\omega - \lambda\,dx = dg$ on $[0,1]$ to find λ.

4-30. If ω is a 1-form on $\mathbf{R}^2 - 0$ such that $d\omega = 0$, prove that

$$\omega = \lambda\,d\theta + dg$$

for some $\lambda \in \mathbf{R}$ and g: $\mathbf{R}^2 - 0 \to \mathbf{R}$. *Hint:* If

$$c_{R,1}{}^*(\omega) = \lambda_R\,dx + d(g_R),$$

show that all numbers λ_R have the same value λ.

4-31. If $\omega \neq 0$, show that there is a chain c such that $\int_c \omega \neq 0$. Use this fact, Stokes' theorem and $\partial^2 = 0$ to prove $d^2 = 0$.

4-32. (a) Let c_1, c_2 be singular 1-cubes in $\mathbf{R}^2$ with $c_1(0) = c_2(0)$ and $c_1(1) = c_2(1)$. Show that there is a singular 2-cube c such that $\partial c = c_1 - c_2 + c_3 - c_4$, where c_3 and c_4 are *degenerate,* that is, $c_3([0,1])$ and $c_4([0,1])$ are points. Conclude that $\int_{c_1} \omega = \int_{c_2} \omega$ if ω is exact. Give a counterexample on $\mathbf{R}^2 - 0$ if ω is merely closed.

(b) If ω is a 1-form on a subset of $\mathbf{R}^2$ and $\int_{c_1} \omega = \int_{c_2} \omega$ for all c_1, c_2 with $c_1(0) = c_2(0)$ and $c_1(1) = c_2(1)$, show that ω is exact. *Hint:* Consider Problems 2-21 and 3-34.

4-33. (*A first course in complex variables.*) If f: $\mathbf{C} \to \mathbf{C}$, define f to be ***differentiable*** at $z_0 \in \mathbf{C}$ if the limit

$$f'(z_0) = \lim_{z \to z_0} \frac{f(z) - f(z_0)}{z - z_0}$$

exists. (This quotient involves two complex numbers and this definition is completely different from the one in Chapter 2.) If f is differentiable at every point z in an open set A and f' is continuous on A, then f is called ***analytic*** on A.

(a) Show that $f(z) = z$ is analytic and $f(z) = \bar{z}$ is not (where $\overline{x + iy} = x - iy$). Show that the sum, product, and quotient of analytic functions are analytic.

(b) If $f = u + iv$ is analytic on A, show that u and v satisfy the *Cauchy-Riemann equations:*

$$\frac{\partial u}{\partial x} = \frac{\partial v}{\partial y} \quad \text{and} \quad \frac{\partial u}{\partial y} = \frac{-\partial v}{\partial x}.$$

Hint: Use the fact that $\lim_{z \to z_0} [f(z) - f(z_0)]/(z - z_0)$ must be the same for $z = z_0 + (x + i \cdot 0)$ and $z = z_0 + (0 + i \cdot y)$ with $x,y \to 0$. (The converse is also true, if u and v are continuously differentiable; this is more difficult to prove.)

(c) Let T: $\mathbf{C} \to \mathbf{C}$ be a linear transformation (where $\mathbf{C}$ is considered as a vector space over $\mathbf{R}$). If the matrix of T with respect to the basis $(1,i)$ is $\begin{pmatrix} a,b \\ c,d \end{pmatrix}$ show that T is multiplication by a comlex number if and only if $a = d$ and $b = -c$. Part (b) shows that an analytic function f: $\mathbf{C} \to \mathbf{C}$, considered as a function f: $\mathbf{R}^2 \to \mathbf{R}^2$, has a derivative $Df(z_0)$ which is multiplication by a complex number. What complex number is this?

(d) Define

$$d(\omega + i\eta) = d\omega + i\, d\eta,$$
$$\int_c \omega + i\eta = \int_c \omega + i \int_c \eta,$$

$$(\omega + i\eta) \wedge (\theta + i\lambda) = \omega \wedge \theta - \eta \wedge \lambda + i(\eta \wedge \theta + \omega \wedge \lambda),$$

and

$$dz = dx + i\, dy.$$

Show that $d(f \cdot dz) = 0$ if and only if f satisfies the Cauchy-Riemann equations.

(e) Prove the *Cauchy Integral Theorem:* If f is analytic on A, then $\int_c f\, dz = 0$ for every closed curve c (singular 1-cube with $c(0) = c(1)$) such that $c = \partial c'$ for some 2-chain c' in A.

(f) Show that if $g(z) = 1/z$, then $g \cdot dz$ [or $(1/z)dz$ in classical notation] equals $i\, d\theta + dh$ for some function h: $\mathbf{C} - 0 \to \mathbf{R}$. Conclude that $\int_{c_{R,n}} (1/z)dz = 2\pi i n$.

(g) If f is analytic on $\{z\colon |z| < 1\}$, use the fact that $g(z) = f(z)/z$ is analytic in $\{z\colon 0 < |z| < 1\}$ to show that

$$\int_{c_{R_1,n}} \frac{f(z)}{z}\, dz = \int_{c_{R_2,n}} \frac{f(z)}{z}\, dz$$

if $0 < R_1$, $R_2 < 1$. Use (f) to evaluate $\lim_{R \to 0} \int_{c_{R,n}} f(z)/z\, dz$ and conclude:

Cauchy Integral Formula: If f is analytic on $\{z\colon |z| < 1\}$ and c is a closed curve in $\{z\colon 0 < |z| < 1\}$ with winding number n around 0, then

$$n \cdot f(0) = \frac{1}{2\pi i} \int_c \frac{f(z)}{z}\, dz.$$

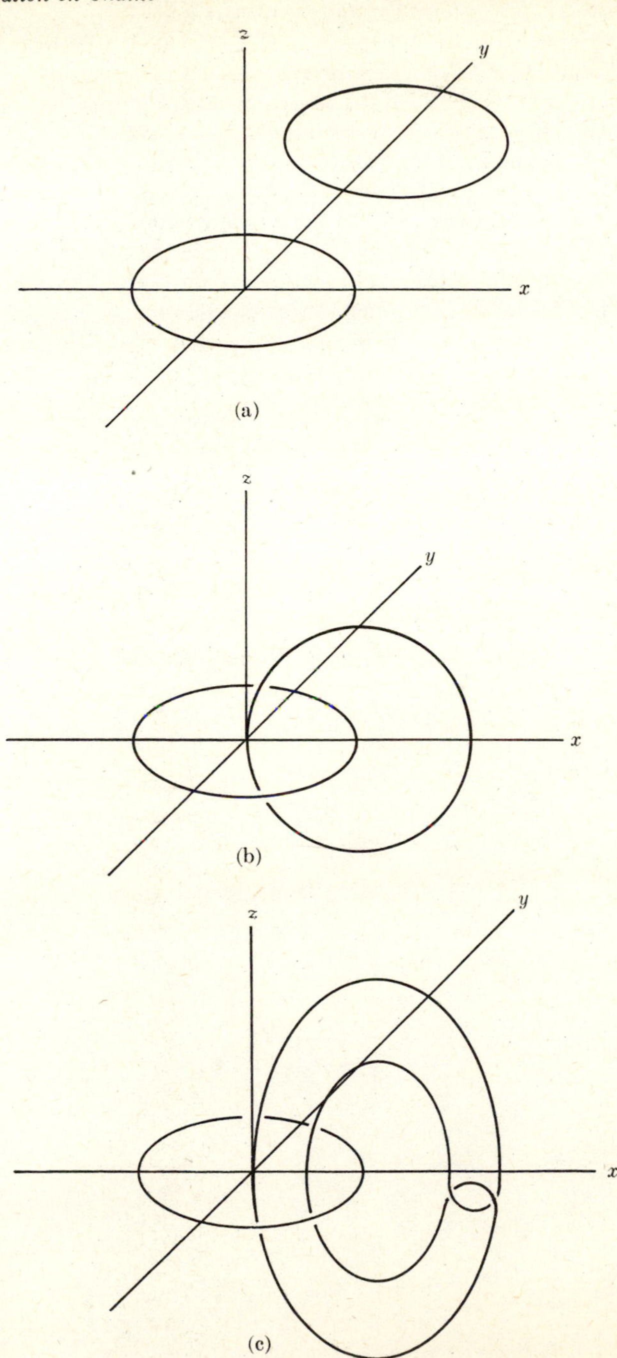

FIGURE 4-6

4-34. If $F\colon [0,1]^2 \to \mathbf{R}^3$ and $s \in [0,1]$ define $F_s\colon [0,1] \to \mathbf{R}^3$ by $F_s(t) = F(s,t)$. If each F_s is a closed curve, F is called a ***homotopy*** between the closed curve F_0 and the closed curve F_1. Suppose F and G are homotopies of closed curves; if for each s the closed curves F_s and G_s do not intersect, the pair (F,G) is called a homotopy between the nonintersecting closed curves F_0, G_0 and F_1, G_1. It is intuitively obvious that there is no such homotopy with F_0, G_0 the pair of curves shown in Figure 4-6 (a), and F_1, G_1 the pair of (b) or (c). The present problem, and Problem 5-33 prove this for (b) but the proof for (c) requires different techniques.

(a) If $f, g\colon [0,1] \to \mathbf{R}^3$ are nonintersecting closed curves define $c_{f,g}\colon [0,1]^2 \to \mathbf{R}^3 - 0$ by

$$c_{f,g}(u,v) = f(u) - g(v).$$

If (F,G) is a homotopy of nonintersecting closed curves define $C_{F,G}\colon [0,1]^3 \to \mathbf{R}^3 - 0$ by

$$C_{F,G}(s,u,v) = c_{F_s,G_s}(u,v) = F(s,u) - G(s,v).$$

Show that

$$\partial C_{F,G} = c_{F_0,G_0} - c_{F_1,G_1}.$$

(b) If ω is a closed 2-form on $\mathbf{R}^3 - 0$ show that

$$\int_{c_{F_0,G_0}} \omega = \int_{c_{F_1,G_1}} \omega.$$

5

Integration on Manifolds

MANIFOLDS

If U and V are open sets in $\mathbf{R}^n$, a differentiable function $h: U \to V$ with a differentiable inverse $h^{-1}: V \to U$ will be called a **diffeomorphism.** ("Differentiable" henceforth means "C^∞".)

A subset M of $\mathbf{R}^n$ is called a **k-dimensional manifold** (in $\mathbf{R}^n$) if for every point $x \in M$ the following condition is satisfied:

(M) There is an open set U containing x, an open set $V \subset \mathbf{R}^n$, and a diffeomorphism $h: U \to V$ such that

$$\begin{aligned} h(U \cap M) &= V \cap (\mathbf{R}^k \times \{0\}) \\ &= \{y \in V: y^{k+1} = \cdots = y^n = 0\}. \end{aligned}$$

In other words, $U \cap M$ is, "up to diffeomorphism," simply $\mathbf{R}^k \times \{0\}$ (see Figure 5-1). The two extreme cases of our definition should be noted: a point in $\mathbf{R}^n$ is a 0-dimensional manifold, and an open subset of $\mathbf{R}^n$ is an n-dimensional manifold.

One common example of an n-dimensional manifold is the

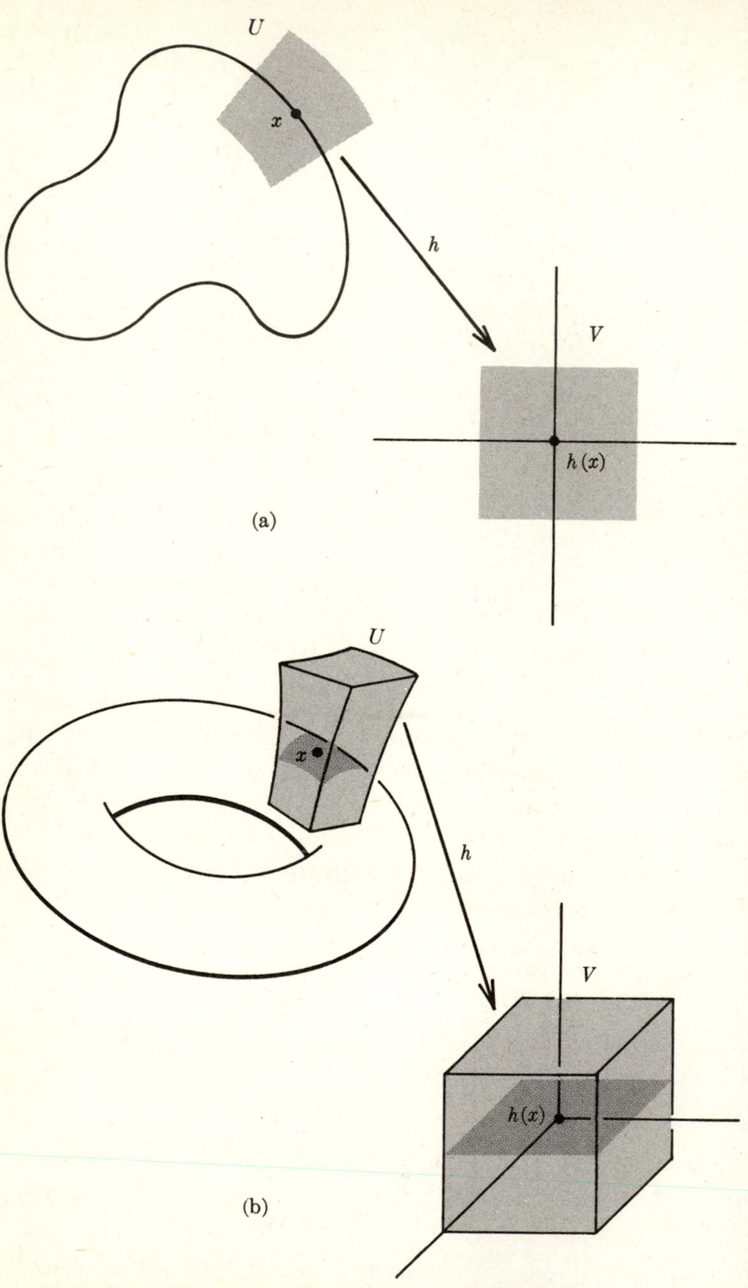

FIGURE 5-1. *A one-dimensional manifold in* $\mathbf{R}^2$ *and a two-dimensional manifold in* $\mathbf{R}^3$.

n-sphere S^n, defined as $\{x \in \mathbf{R}^{n+1}: |x| = 1\}$. We leave it as an exercise for the reader to prove that condition (M) is satisfied. If you are unwilling to trouble yourself with the details, you may instead use the following theorem, which provides many examples of manifolds (note that $S^n = g^{-1}(0)$, where $g: \mathbf{R}^{n+1} \to \mathbf{R}$ is defined by $g(x) = |x|^2 - 1$).

5-1 Theorem. *Let $A \subset \mathbf{R}^n$ be open and let $g: A \to \mathbf{R}^p$ be a differentiable function such that $g'(x)$ has rank p whenever $g(x) = 0$. Then $g^{-1}(0)$ is an $(n - p)$-dimensional manifold in $\mathbf{R}^n$.*

Proof. This follows immediately from Theorem 2-13. ▌

There is an alternative characterization of manifolds which is very important.

5-2 Theorem. *A subset M of $\mathbf{R}^n$ is a k-dimensional manifold if and only if for each point $x \in M$ the following "coordinate condition" is satisfied:*

(C) There is an open set U containing x, an open set $W \subset \mathbf{R}^k$, and a 1-1 differentiable function $f: W \to \mathbf{R}^n$ such that

(1) *$f(W) = M \cap U$,*
(2) *$f'(y)$ has rank k for each $y \in W$,*
(3) *$f^{-1}: f(W) \to W$ is continuous.*

[Such a function f is called a **coordinate system** around x (see Figure 5-2).]

Proof. If M is a k-dimensional manifold in $\mathbf{R}^n$, choose $h: U \to V$ satisfying (M). Let $W = \{a \in \mathbf{R}^k: (a,0) \in h(M)\}$ and define $f: W \to \mathbf{R}^n$ by $f(a) = h^{-1}(a,0)$. Clearly $f(W) = M \cap U$ and f^{-1} is continuous. If $H: U \to \mathbf{R}^k$ is $H(z) = (h^1(z), \ldots, h^k(z))$, then $H(f(y)) = y$ for all $y \in W$; therefore $H'(f(y)) \cdot f'(y) = I$ and $f'(y)$ must have rank k.

Suppose, conversely, that $f: W \to \mathbf{R}^n$ satisfies condition (C). Let $x = f(y)$. Assume that the matrix $(D_j f^i(y))$, $1 \leq i, j \leq k$ has a non-zero determinant. Define $g: W \times \mathbf{R}^{n-k} \to \mathbf{R}^n$ by

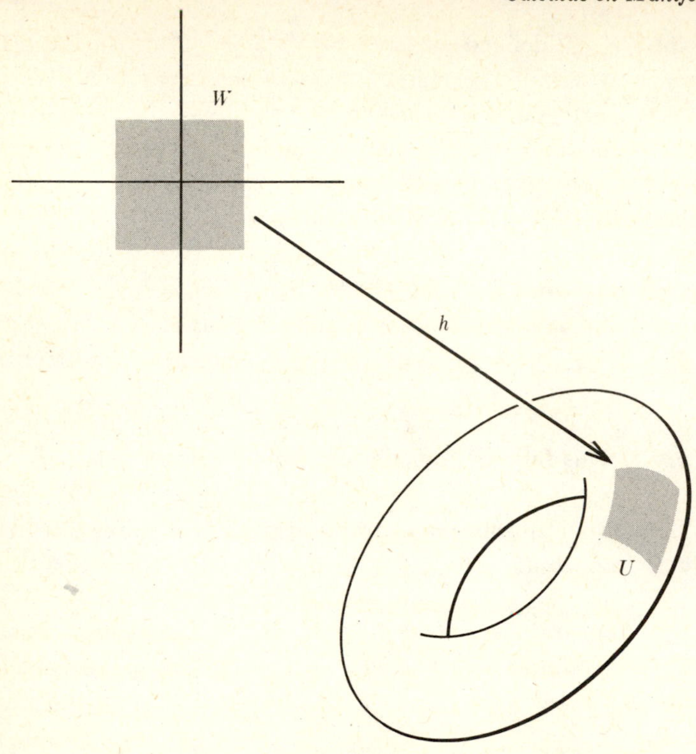

FIGURE 5-2

$g(a,b) = f(a) + (0,b)$. Then $\det g'(a,b) = \det (D_j f^i(a))$, so $\det g'(y,0) \neq 0$. By Theorem 2-11 there is an open set V_1' containing $(y,0)$ and an open set V_2' containing $g(y,0) = x$ such that $g\colon V_1' \to V_2'$ has a differentiable inverse $h\colon V_2' \to V_1'$. Since f^{-1} is continuous, $\{f(a)\colon (a,0) \in V_1'\} = U \cap f(W)$ for some open set U. Let $V_2 = V_2' \cap U$ and $V_1 = g^{-1}(V_2)$. Then $V_2 \cap M$ is exactly $\{f(a)\colon (a,0) \in V_1\} = \{g(a,0)\colon (a,0) \in V_1\}$, so

$$\begin{aligned} h(V_2 \cap M) = g^{-1}(V_2 \cap M) &= g^{-1}(\{g(a,0)\colon (a,0) \in V_1\}) \\ &= V_1 \cap (\mathbf{R}^k \times \{0\}). \quad \blacksquare \end{aligned}$$

One consequence of the proof of Theorem 5-2 should be noted. If $f_1\colon W_1 \to \mathbf{R}^n$ and $f_2\colon W_2 \to \mathbf{R}^n$ are two coordinate systems, then

$$f_2^{-1} \circ f_1 \colon f_1^{-1}(f_2(W_2)) \to \mathbf{R}^k$$

is differentiable with non-singular Jacobian. If fact, $f_2^{-1}(y)$ consists of the first k components of $h(y)$.

The **half-space** $\mathbf{H}^k \subset \mathbf{R}^k$ is defined as $\{x \in \mathbf{R}^k \colon x^k \geq 0\}$. A subset M of $\mathbf{R}^n$ is a **k-dimensional manifold-with-boundary** (Figure 5-3) if for every point $x \in M$ either condition (M) or the following condition is satisfied:

(M') There is an open set U containing x, an open set $V \subset \mathbf{R}^n$, and a diffeomorphism $h \colon U \to V$ such that

$$\begin{aligned} h(U \cap M) &= V \cap (\mathbf{H}^k \times \{0\}) \\ &= \{y \in V \colon y^k \geq 0 \text{ and } y^{k+1} = \cdots = y^n = 0\} \end{aligned}$$

and $h(x)$ has kth component $= 0$.

It is important to note that conditions (M) and (M') cannot both hold for the same x. In fact, if $h_1 \colon U_1 \to V_1$ and $h_2 \colon U_2 \to V_2$ satisfied (M) and (M'), respectively, then $h_2 \circ h_1^{-1}$ would be a differentiable map that takes an open set in $\mathbf{R}^k$, containing $h(x)$, into a subset of $\mathbf{H}^k$ which is not open in $\mathbf{R}^k$. Since $\det (h_2 \circ h_1^{-1})' \neq 0$, this contradicts Problem 2-36. The set of all points $x \in M$ for which condition M' is satisfied is called the **boundary** of M and denoted ∂M. This

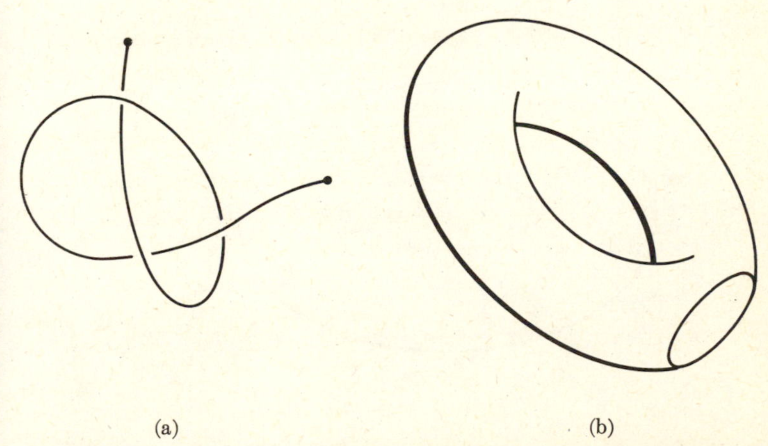

FIGURE 5-3. *A one-dimensional and a two-dimensional manifold-with-boundary in* $\mathbf{R}^3$.

must not be confused with the boundary of a set, as defined in Chapter 1 (see Problems 5-3 and 5-8).

Problems. **5-1.** If M is a k-dimensional manifold-with-boundary, prove that ∂M is a $(k-1)$-dimensional manifold and $M - \partial M$ is a k-dimensional manifold.

5-2. Find a counterexample to Theorem 5-2 if condition (3) is omitted. *Hint:* Wrap an open interval into a figure six.

5-3. (a) Let $A \subset \mathbf{R}^n$ be an open set such that boundary A is an $(n-1)$-dimensional manifold. Show that $N = A \cup$ boundary A is an n-dimensional manifold-with-boundary. (It is well to bear in mind the following example: if $A = \{x \in \mathbf{R}^n: |x| < 1 \text{ or } 1 < |x| < 2\}$ then $N = A \cup$ boundary A is a manifold-with-boundary, but $\partial N \neq$ boundary A.)

(b) Prove a similar assertion for an open subset of an n-dimensional manifold.

5-4. Prove a partial converse of Theorem 5-1: If $M \subset \mathbf{R}^n$ is a k-dimensional manifold and $x \in M$, then there is an open set $A \subset \mathbf{R}^n$ containing x and a differentiable function $g: A \to \mathbf{R}^{n-k}$ such that $A \cap M = g^{-1}(0)$ and $g'(y)$ has rank $n - k$ when $g(y) = 0$.

5-5. Prove that a k-dimensional (vector) subspace of $\mathbf{R}^n$ is a k-dimensional manifold.

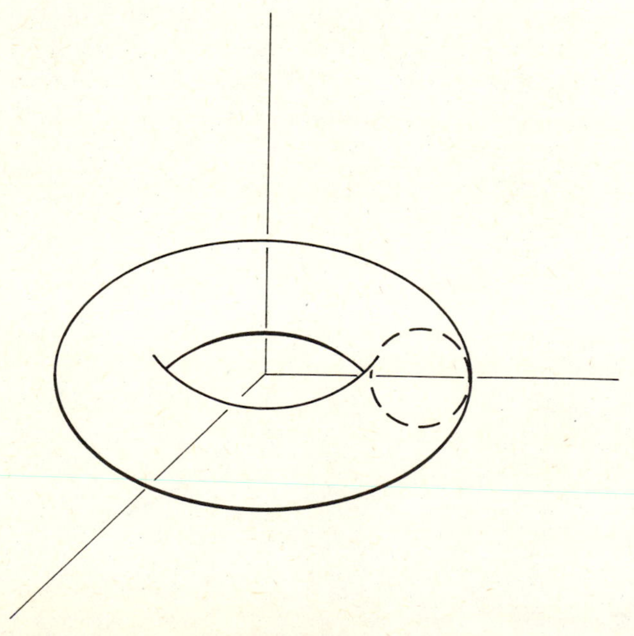

FIGURE 5-4

5-6. If $f: \mathbf{R}^n \to \mathbf{R}^m$, the **graph** of f is $\{(x,y): y = f(x)\}$. Show that the graph of f is an n-dimensional manifold if and only if f is differentiable.

5-7. Let $\mathbf{K}^n = \{x \in \mathbf{R}^n: x^1 = 0 \text{ and } x^2, \ldots, x^{n-1} > 0\}$. If $M \subset \mathbf{K}^n$ is a k-dimensional manifold and N is obtained by revolving M around the axis $x^1 = \cdots = x^{n-1} = 0$, show that N is a $(k+1)$-dimensional manifold. Example: the torus (Figure 5-4).

5-8. (a) If M is a k-dimensional manifold in $\mathbf{R}^n$ and $k < n$, show that M has measure 0.

(b) If M is a closed n-dimensional manifold-with-boundary in $\mathbf{R}^n$, show that the boundary of M is ∂M. Give a counterexample if M is not closed.

(c) If M is a compact n-dimensional manifold-with-boundary in $\mathbf{R}^n$, show that M is Jordan-measurable.

FIELDS AND FORMS ON MANIFOLDS

Let M be a k-dimensional manifold in $\mathbf{R}^n$ and let $f: W \to \mathbf{R}^n$ be a coordinate system around $x = f(a)$. Since $f'(a)$ has rank k, the linear transformation $f_*: \mathbf{R}^k{}_a \to \mathbf{R}^n{}_x$ is 1-1, and $f_*(\mathbf{R}^k{}_a)$ is a k-dimensional subspace of $\mathbf{R}^n{}_x$. If $g: V \to \mathbf{R}^n$ is another coordinate system, with $x = g(b)$, then

$$g_*(\mathbf{R}^k{}_b) = f_*(f^{-1} \circ g)_*(\mathbf{R}^k{}_b) = f_*(\mathbf{R}^k{}_a).$$

Thus the k-dimensional subspace $f_*(\mathbf{R}^k{}_a)$ does not depend on the coordinate system f. This subspace is denoted M_x, and is called the **tangent space** of M at x (see Figure 5-5). In later sections we will use the fact that there is a natural inner product T_x on M_x, induced by that on $\mathbf{R}^n{}_x$: if $v,w \in M_x$ define $T_x(v,w) = \langle v,w \rangle_x$.

Suppose that A is an open set containing M, and F is a differentiable vector field on A such that $F(x) \in M_x$ for each $x \in M$. If $f: W \to \mathbf{R}^n$ is a coordinate system, there is a unique (differentiable) vector field G on W such that $f_*(G(a)) = F(f(a))$ for each $a \in W$. We can also consider a function F which merely assigns a vector $F(x) \in M_x$ for each $x \in M$; such a function is called a **vector field on** M. There is still a unique vector field G on W such that $f_*(G(a)) = F(f(a))$ for $a \in W$; we *define* F to be differentiable if G is differentiable. Note that our definition does not depend on the coordinate

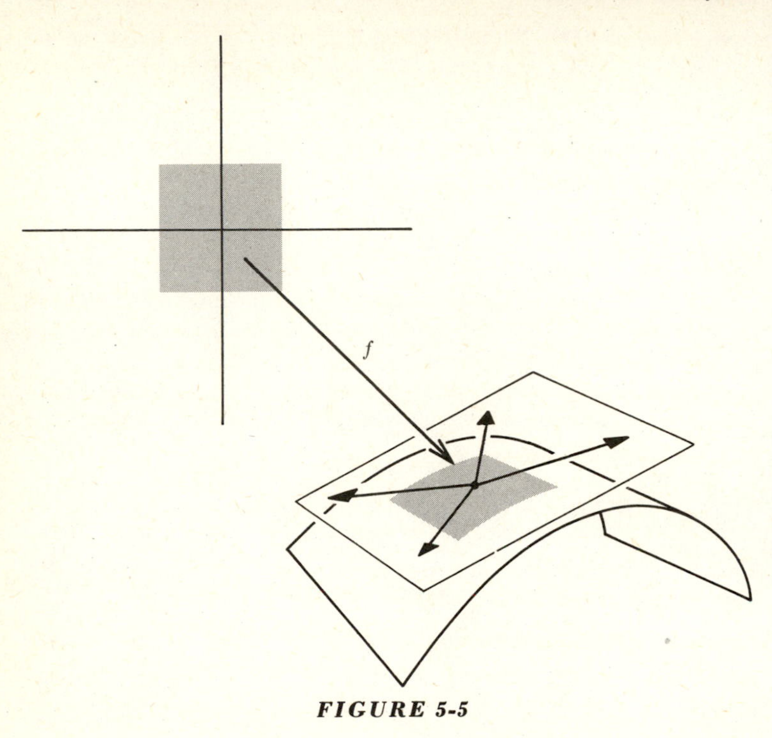

FIGURE 5-5

system chosen: if $g: V \to \mathbf{R}^n$ and $g_*(H(b)) = F(g(b))$ for all $b \in V$, then the component functions of $H(b)$ must equal the component functions of $G(f^{-1}(g(b)))$, so H is differentiable if G is.

Precisely the same considerations hold for forms. A function ω which assigns $\omega(x) \in \Lambda^p(M_x)$ for each $x \in M$ is called a **p-form on** M. If $f: W \to \mathbf{R}^n$ is a coordinate system, then $f^*\omega$ is a p-form on W; we *define* ω to be differentiable if $f^*\omega$ is. A p-form ω on M can be written as

$$\omega = \sum_{i_1 < \cdots < i_p} \omega_{i_1, \ldots, i_p} \, dx^{i_1} \wedge \cdots \wedge dx^{i_p}.$$

Here the functions $\omega_{i_1, \ldots, i_p}$ are defined only on M. The definition of $d\omega$ given previously would make no sense here, since $D_j(\omega_{i_1, \ldots, i_p})$ has no meaning. Nevertheless, there is a reasonable way of defining $d\omega$.

5-3 Theorem. *There is a unique $(p + 1)$-form $d\omega$ on M such that for every coordinate system $f: W \to \mathbf{R}^n$ we have*

$$f^*(d\omega) = d(f^*\omega).$$

Proof. If $f: W \to \mathbf{R}^n$ is a coordinate system with $x = f(a)$ and $v_1, \ldots, v_{p+1} \in M_x$, there are unique $w_1, \ldots, w_{p+1}$ in $\mathbf{R}^k{}_a$ such that $f_*(w_i) = v_i$. Define $d\omega(x)(v_1, \ldots, v_{p+1}) = d(f^*\omega)(a)(w_1, \ldots, w_{p+1})$. One can check that this definition of $d\omega(x)$ does not depend on the coordinate system f, so that $d\omega$ is well-defined. Moreover, it is clear that $d\omega$ has to be defined this way, so $d\omega$ is unique. ∎

It is often necessary to choose an orientation μ_x for each tangent space M_x of a manifold M. Such choices are called **consistent** (Figure 5-6) provided that for every coordinate

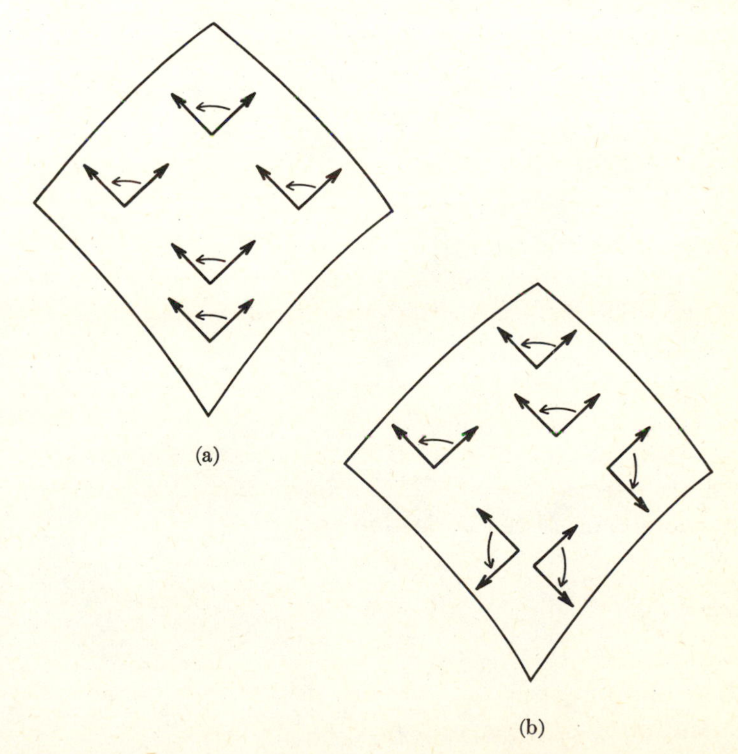

FIGURE 5-6. *(a) Consistent and (b) inconsistent choices of orientations.*

system $f\colon W \to \mathbf{R}^n$ and $a,b \in W$ the relation

$$[f_*((e_1)_a), \ldots ,f_*((e_k)_a)] = \mu_{f(a)}$$

holds if and only if

$$[f_*((e_1)_b), \ldots ,f_*((e_k)_b)] = \mu_{f(b)}.$$

Suppose orientations μ_x have been chosen consistently. If $f\colon W \to \mathbf{R}^n$ is a coordinate system such that

$$[f_*((e_1)_a), \ldots , f_*((e_k)_a)] = \mu_{f(a)}$$

for one, and hence for every $a \in W$, then f is called **orientation-preserving.** If f is *not* orientation-preserving and $T\colon \mathbf{R}^k \to \mathbf{R}^k$ is a linear transformation with $\det T = -1$, then $f \circ T$ *is* orientation-preserving. Therefore there is an orientation-preserving coordinate system around each point. If f and g are orientation-preserving and $x = f(a) = g(b)$, then the relation

$$[f_*((e_1)_a), \ldots ,f_*((e_k)_a)] = \mu_x = [g_*((e_1)_b), \ldots ,g_*((e_k)_b)]$$

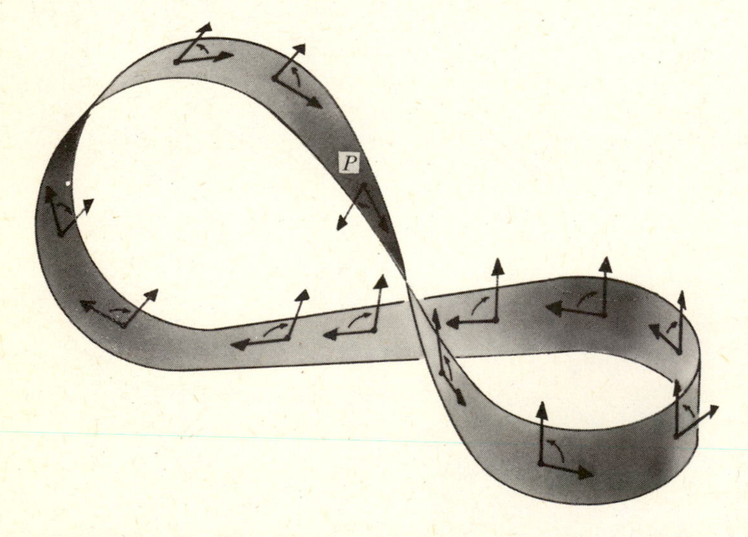

FIGURE 5-7. *The Möbius strip, a non-orientable manifold. A basis begins at P, moves to the right and around, and comes back to P with the wrong orientation.*

implies that

$$[(g^{-1} \circ f)_*((e_1)_a), \ldots, (g^{-1} \circ f)_*((e_k)_a)] = [(e_1)_b, \ldots, (e_k)_b],$$

so that $\det (g^{-1} \circ f)' > 0$, an important fact to remember.

A manifold for which orientations μ_x can be chosen consistently is called **orientable**, and a particular choice of the μ_x is called an **orientation** μ of M. A manifold together with an orientation μ is called an **oriented** manifold. The classical example of a non-orientable manifold is the Möbius strip. A model can be made by gluing together the ends of a strip of paper which has been given a half twist (Figure 5-7).

Our definitions of vector fields, forms, and orientations can be made for manifolds-with-boundary also. If M is a k-dimensional manifold-with-boundary and $x \in \partial M$, then $(\partial M)_x$ is a $(k - 1)$-dimensional subspace of the k-dimensional vector space M_x. Thus there are exactly two unit vectors in M_x which are perpendicular to $(\partial M)_x$; they can be distinguished as follows (Figure 5-8). If $f: W \to \mathbf{R}^n$ is a coordinate system with $W \subset H^k$ and $f(0) = x$, then only one of these unit vectors is $f_*(v_0)$ for some v_0 with $v^k < 0$. This unit vector is called the **outward unit normal** $n(x)$; it is not hard to check that this definition does not depend on the coordinate system f.

Suppose that μ is an orientation of a k-dimensional manifold-with-boundary M. If $x \in \partial M$, choose $v_1, \ldots, v_{k-1} \in (\partial M)_x$ so that $[n(x), v_1, \ldots, v_{k-1}] = \mu_x$. If it is also true that $[n(x), w_1, \ldots, w_{k-1}] = \mu_x$, then both $[v_1, \ldots, v_{k-1}]$ and $[w_1, \ldots, w_{k-1}]$ are the same orientation for $(\partial M)_x$. This orientation is denoted $(\partial\mu)_x$. It is easy to see that the orientations $(\partial\mu)_x$, for $x \in \partial M$, are consistent on ∂M. Thus if M is orientable, ∂M is also orientable, and an orientation μ for M determines an orientation $\partial\mu$ for ∂M, called the **induced orientation.** If we apply these definitions to $\mathbf{H}^k$ with the usual orientation, we find that the induced orientation on $\mathbf{R}^{k-1} = \{x \in \mathbf{H}^k: x^k = 0\}$ is $(-1)^k$ times the usual orientation. The reason for such a choice will become clear in the next section.

If M is an *oriented* $(n - 1)$-dimensional manifold in $\mathbf{R}^n$, a substitute for outward unit normal vectors can be defined,

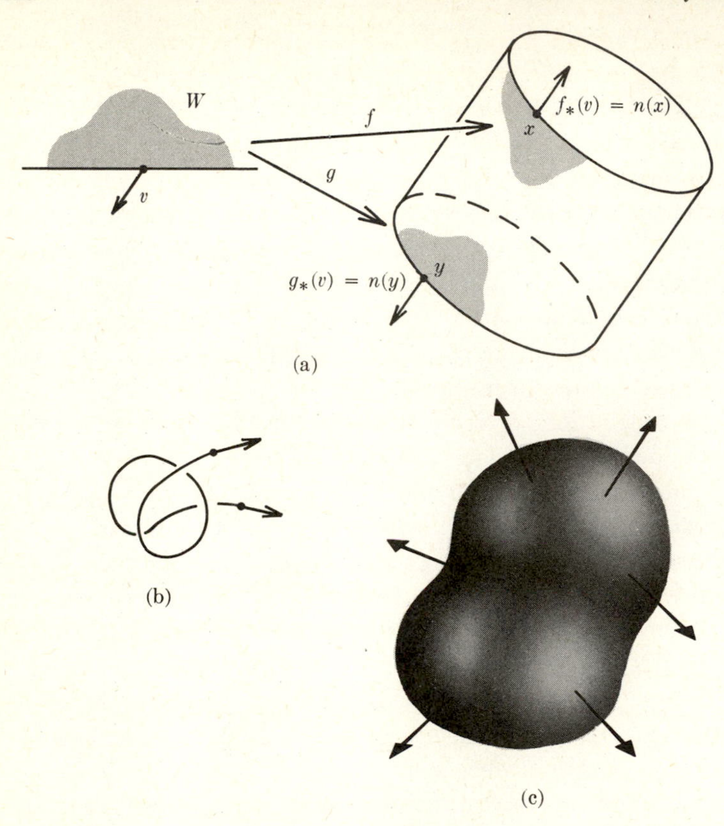

FIGURE 5-8. *Some outward unit normal vectors of manifolds-with-boundary in* $\mathbf{R}^3$.

even though M is not necessarily the boundary of an n-dimensional manifold. If $[v_1, \ldots, v_{n-1}] = \mu_x$, we choose $n(x)$ in $\mathbf{R}^n{}_x$ so that $n(x)$ is a unit vector perpendicular to M_x and $[n(x), v_1, \ldots, v_{n-1}]$ is the usual orientation of $\mathbf{R}^n{}_x$. We still call $n(x)$ the outward unit normal to M (determined by μ). The vectors $n(x)$ vary continuously on M, in an obvious sense. Conversely, if a continuous family of unit normal vectors $n(x)$ is defined on all of M, then we can determine an orientation of M. This shows that such a continuous choice of normal vectors is impossible on the Möbius strip. In the paper model of the Möbius strip the two sides of the paper (which has thickness) may be thought of as the end points of the unit normal vectors in both directions. The impossibility of choosing normal vectors continuously is reflected by the famous property of the paper model. The paper model is one-sided (if you start to paint it on one side you end up painting it all over); in other words, choosing $n(x)$ arbitrarily at one point, and then by the continuity requirement at other points, eventually forces the opposite choice for $n(x)$ at the initial point.

Problems. **5-9.** Show that M_x consists of the tangent vectors at t of curves c in M with $c(t) = x$.

5-10. Suppose $\mathcal{C}$ is a collection of coordinate systems for M such that (1) For each $x \in M$ there is $f \in \mathcal{C}$ which is a coordinate system around x; (2) if $f,g \in \mathcal{C}$, then $\det (f^{-1} \circ g)' > 0$. Show that there is a unique orientation of M such that f is orientation-preserving if $f \in \mathcal{C}$.

5-11. If M is an n-dimensional manifold-with-boundary in $\mathbf{R}^n$, define μ_x as the usual orientation of $M_x = \mathbf{R}^n{}_x$ (the orientation μ so defined is the **usual orientation** of M). If $x \in \partial M$, show that the two definitions of $n(x)$ given above agree.

5-12. (a) If F is a differentiable vector field on $M \subset \mathbf{R}^n$, show that there is an open set $A \supset M$ and a differentiable vector field $\tilde{F}$ on A with $\tilde{F}(x) = F(x)$ for $x \in M$. *Hint:* Do this locally and use partitions of unity.

(b) If M is closed, show that we can choose $A = \mathbf{R}^n$.

5-13. Let $g\colon A \to \mathbf{R}^p$ be as in Theorem 5-1.

(a) If $x \in M = g^{-1}(0)$, let $h\colon U \to \mathbf{R}^n$ be the essentially unique diffeomorphism such that $g \circ h(y) = (y^{n-p+1}, \ldots, y^n)$ and $h(0) = x$. Define $f\colon \mathbf{R}^{n-p} \to \mathbf{R}^n$ by $f(a) = h(0,a)$. Show that f_* is 1-1 so that the $n - p$ vectors $f_*((e_1)_0), \ldots, f_*((e_{n-p})_0)$ are linearly independent.

(b) Show that orientations μ_x can be defined consistently, so that M is orientable.

(c) If $p = 1$, show that the components of the outward normal at x are some multiple of $D_1g(x), \ldots, D_ng(x)$.

5-14. If $M \subset \mathbf{R}^n$ is an orientable $(n - 1)$-dimensional manifold, show that there is an open set $A \subset \mathbf{R}^n$ and a differentiable $g\colon A \to \mathbf{R}^1$ so that $M = g^{-1}(0)$ and $g'(x)$ has rank 1 for $x \in M$. *Hint:* Problem 5-4 does this locally. Use the orientation to choose consistent local solutions and use partitions of unity.

5-15. Let M be an $(n - 1)$-dimensional manifold in $\mathbf{R}^n$. Let $M(\varepsilon)$ be the set of end points of normal vectors (in both directions) of length ε and suppose ε is small enough so that $M(\varepsilon)$ is also an

$(n-1)$-dimensional manifold. Show that $M(\varepsilon)$ is orientable (even if M is not). What is $M(\varepsilon)$ if M is the Möbius strip?

5-16. Let $g\colon A \to \mathbf{R}^p$ be as in Theorem 5-1. If $f\colon \mathbf{R}^n \to \mathbf{R}$ is differentiable and the maximum (or minimum) of f on $g^{-1}(0)$ occurs at a, show that there are $\lambda_1, \ldots, \lambda_p \in \mathbf{R}$, such that

$$(1)\quad D_jf(a) = \sum_{i=1}^{n} \lambda_i D_j g^i(a) \qquad j = 1, \ldots, n.$$

Hint: This equation can be written $df(a) = \sum_{i=1}^{n} \lambda_i dg^i(a)$ and is obvious if $g(x) = (x^{n-p+1}, \ldots, x^n)$.

The maximum of f on $g^{-1}(0)$ is sometimes called the maximum of f subject to the **constraints** $g^i = 0$. One can attempt to find a by solving the system of equations (1). In particular, if $g\colon A \to \mathbf{R}$, we must solve $n + 1$ equations

$$D_jf(a) = \lambda D_j g(a),$$
$$g(a) = 0,$$

in $n + 1$ unknowns $a^1, \ldots, a^n, \lambda$, which is often very simple if we leave the equation $g(a) = 0$ for last. This is **Lagrange's method,** and the useful but irrelevant λ is called a **Lagrangian multiplier.** The following problem gives a nice theoretical use for Lagrangian multipliers.

5-17. (a) Let $T\colon \mathbf{R}^n \to \mathbf{R}^n$ be self-adjoint with matrix $A = (a_{ij})$, so that $a_{ij} = a_{ji}$. If $f(x) = \langle Tx, x\rangle = \sum a_{ij}x^ix^j$, show that $D_kf(x) = 2\sum_{j=1}^{n} a_{kj}x^j$. By considering the maximum of $\langle Tx, x\rangle$ on S^{n-1} show that there is $x \in S^{n-1}$ and $\lambda \in \mathbf{R}$ with $Tx = \lambda x$.

(b) If $V = \{y \in \mathbf{R}^n\colon \langle x, y\rangle = 0\}$, show that $T(V) \subset V$ and $T\colon V \to V$ is self-adjoint.

(c) Show that T has a basis of eigenvectors.

STOKES' THEOREM ON MANIFOLDS

If ω is a p-form on a k-dimensional manifold-with-boundary M and c is a singular p-cube in M, we define

$$\int_c \omega = \int_{[0,1]^p} c^*\omega$$

precisely as before; integrals over p-chains are also defined as before. In the case $p = k$ it may happen that there is an open set $W \supset [0,1]^k$ and a coordinate system $f\colon W \to \mathbf{R}^n$ such that $c(x) = f(x)$ for $x \in [0,1]^k$; a k-cube in M will always be understood to be of this type. If M is oriented, the singular k-cube c is called **orientation-preserving** if f is.

5-4 *Theorem.* *If* $c_1,c_2\colon [0,1]^k \to M$ *are two* orientation-preserving *singular k-cubes in the oriented k-dimensional manifold M and ω is a k-form on M such that $\omega = 0$ outside of* $c_1([0,1]^k) \cap c_2([0,1]^k)$, *then*

$$\int_{c_1} \omega = \int_{c_2} \omega.$$

Proof. We have

$$\int_{c_1} \omega = \int_{[0,1]^k} c_1{}^*(\omega) = \int_{[0,1]^k} (c_2^{-1} \circ c_1)^* c_2{}^*(\omega).$$

(Here $c_2^{-1} \circ c_1$ is defined only on a subset of $[0,1]^k$ and the second equality depends on the fact that $\omega = 0$ outside of $c_1([0,1]^k) \cap c_2([0,1]^k)$.) It therefore suffices to show that

$$\int_{[0,1]^k} (c_2^{-1} \circ c_1)^* c_2{}^*(\omega) = \int_{[0,1]^k} c_2{}^*(\omega) = \int_{c_2} \omega.$$

If $c_2{}^*(\omega) = f\,dx^1 \wedge \cdots \wedge dx^k$ and $c_2{}^{-1} \circ c_1$ is denoted by g, then by Theorem 4-9 we have

$$\begin{aligned}(c_2{}^{-1} \circ c_1)^* c_2{}^*(\omega) &= g^*(f\,dx^1 \wedge \cdots \wedge dx^k)\\ &= (f \circ g) \cdot \det g' \cdot dx^1 \wedge \cdots \wedge dx^k\\ &= (f \circ g) \cdot |\det g'| \cdot dx^1 \wedge \cdots \wedge dx^k,\end{aligned}$$

since $\det g' = \det(c_2{}^{-1} \circ c_1)' > 0$. The result now follows from Theorem 3-13. ∎

The last equation in this proof should help explain why we have had to be so careful about orientations.

Let ω be a k-form on an oriented k-dimensional manifold M. If there is an orientation-preserving singular k-cube c in M such that $\omega = 0$ outside of $c([0,1]^k)$, we define

$$\int_M \omega = \int_c \omega.$$

Theorem 5-4 shows $\int_M \omega$ does not depend on the choice of c.

Suppose now that ω is an arbitrary k-form on M. There is an open cover $\mathcal{O}$ of M such that for each $U \in \mathcal{O}$ there is an orientation-preserving singular k-cube c with $U \subset c([0,1]^k)$. Let Φ be a partition of unity for M subordinate to this cover. We define

$$\int_M \omega = \sum_{\varphi \in \Phi} \int_M \varphi \cdot \omega$$

provided the sum converges as described in the discussion preceding Theorem 3-12 (this is certainly true if M is compact). An argument similar to that in Theorem 3-12 shows that $\int_M \omega$ does not depend on the cover $\mathcal{O}$ or on Φ.

All our definitions could have been given for a k-dimensional manifold-with-boundary M with orientation μ. Let ∂M have the induced orientation $\partial\mu$. Let c be an orientation-preserving k-cube in M such that $c_{(k,0)}$ lies in ∂M and is the only face which has any interior points in ∂M. As the remarks after the definition of $\partial\mu$ show, $c_{(k,0)}$ is orientation-preserving if k is even, but not if k is odd. Thus, if ω is a $(k-1)$-form on M which is 0 outside of $c([0,1]^k)$, we have

$$\int_{c_{(k,0)}} \omega = (-1)^k \int_{\partial M} \omega.$$

On the other hand, $c_{(k,0)}$ appears with coefficient $(-1)^k$ in ∂c. Therefore

$$\int_{\partial c} \omega = \int_{(-1)^k c_{(k,0)}} \omega = (-1)^k \int_{c_{(k,0)}} \omega = \int_{\partial M} \omega.$$

Our choice of $\partial\mu$ was made to eliminate any minus signs in this equation, and in the following theorem.

5-5 Theorem (Stokes' Theorem). *If M is a compact oriented k-dimensional manifold-with-boundary and ω is a $(k-1)$-form on M, then*

$$\int_M d\omega = \int_{\partial M} \omega.$$

(Here ∂M is given the induced orientation.)

Proof. Suppose first that there is an orientation-preserving singular k-cube in $M - \partial M$ such that $\omega = 0$ outside of

$c([0,1]^k)$. By Theorem 4-13 and the definition of $d\omega$ we have

$$\int_c d\omega = \int_{[0,1]^k} c^*(d\omega) = \int_{[0,1]^k} d(c^*\omega) = \int_{\partial I^k} c^*\omega = \int_{\partial c} \omega.$$

Then

$$\int_M d\omega = \int_c d\omega = \int_{\partial c} \omega = 0,$$

since $\omega = 0$ on ∂c. On the other hand, $\int_{\partial M} \omega = 0$ since $\omega = 0$ on ∂M.

Suppose next that there is an orientation-preserving singular k-cube in M such that $c_{(k,0)}$ is the only face in ∂M, and $\omega = 0$ outside of $c([0,1])^k$. Then

$$\int_M d\omega = \int_c d\omega = \int_{\partial c} \omega = \int_{\partial M} \omega.$$

Now consider the general case. There is an open cover $\mathcal{O}$ of M and a partition of unity Φ for M subordinate to $\mathcal{O}$ such that for each $\varphi \in \Phi$ the form $\varphi \cdot \omega$ is of one of the two sorts already considered. We have

$$0 = d(1) = d\left(\sum_{\varphi \in \Phi} \varphi\right) = \sum_{\varphi \in \Phi} d\varphi,$$

so that

$$\sum_{\varphi \in \Phi} d\varphi \wedge \omega = 0.$$

Since M is compact, this is a finite sum and we have

$$\sum_{\varphi \in \Phi} \int_M d\varphi \wedge \omega = 0.$$

Therefore

$$\begin{aligned}
\int_M d\omega &= \sum_{\varphi \in \Phi} \int_M \varphi \cdot d\omega = \sum_{\varphi \in \Phi} \int_M d\varphi \wedge \omega + \varphi \cdot d\omega \\
&= \sum_{\varphi \in \Phi} \int_M d(\varphi \cdot \omega) = \sum_{\varphi \in \Phi} \int_{\partial M} \varphi \cdot \omega \\
&= \int_{\partial M} \omega. \quad \blacksquare
\end{aligned}$$

Problems. **5-18.** If M is an n-dimensional manifold (or manifold-with-boundary) in $\mathbf{R}^n$, with the usual orientation, show that

$\int_M f\, dx^1 \wedge \cdots \wedge dx^n$, as defined in this section, is the same as $\int_M f$, as defined in Chapter 3.

5-19. (a) Show that Theorem 5-5 is false if M is not compact. *Hint:* If M is a manifold-with-boundary for which 5-5 holds, then $M - \partial M$ is also a manifold-with-boundary (with empty boundary).

(b) Show that Theorem 5-5 holds for noncompact M provided that ω vanishes outside of a compact subset of M.

5-20. If ω is a $(k-1)$-form on a compact k-dimensional manifold M, prove that $\int_M d\omega = 0$. Give a counterexample if M is not compact.

5-21. An **absolute k-tensor** on V is a function $\eta\colon V^k \to \mathbf{R}$ of the form $|\omega|$ for $\omega \in \Lambda^k(V)$. An **absolute k-form** on M is a function η such that $\eta(x)$ is an absolute k-tensor on M_x. Show that $\int_M \eta$ can be defined, even if M is not orientable.

5-22. If $M_1 \subset \mathbf{R}^n$ is an n-dimensional manifold-with-boundary and $M_2 \subset M_1 - \partial M_1$ is an n-dimensional manifold-with-boundary, and M_1, M_2 are compact, prove that

$$\int_{\partial M_1} \omega = \int_{\partial M_2} \omega,$$

where ω is an $(n-1)$-form on M_1, and ∂M_1 and ∂M_2 have the orientations induced by the usual orientations of M_1 and M_2. *Hint:* Find a manifold-with-boundary M such that $\partial M = \partial M_1 \cup \partial M_2$ and such that the induced orientation on ∂M agrees with that for ∂M_1 on ∂M_1 and is the negative of that for ∂M_2 on ∂M_2.

THE VOLUME ELEMENT

Let M be a k-dimensional manifold (or manifold-with-boundary) in $\mathbf{R}^n$, with an orientation μ. If $x \in M$, then μ_x and the inner product T_x we defined previously determine a volume element $\omega(x) \in \Lambda^k(M_x)$. We therefore obtain a nowhere-zero k-form ω on M, which is called the **volume element** on M (determined by μ) and denoted dV, even though it is not generally the differential of a $(k-1)$-form. The **volume** of M is defined as $\int_M dV$, provided this integral exists, which is certainly the case if M is compact. "Volume" is usually called **length** or **surface area** for one- and two-dimensional manifolds, and dV is denoted ds (the "element of length") or dA [or dS] (the "element of [surface] area").

A concrete case of interest to us is the volume element of an oriented surface (two-dimensional manifold) M in $\mathbf{R}^3$. Let $n(x)$ be the unit outward normal at $x \in M$. If $\omega \in \Lambda^2(M_x)$ is defined by

$$\omega(v,w) = \det \begin{pmatrix} v \\ w \\ n(x) \end{pmatrix},$$

then $\omega(v,w) = 1$ if v and w are an orthonormal basis of M_x with $[v,w] = \mu_x$. Thus $dA = \omega$. On the other hand, $\omega(v,w) = \langle v \times w, n(x) \rangle$ by definition of $v \times w$. Thus we have

$$dA(v,w) = \langle v \times w, n(x) \rangle.$$

Since $v \times w$ is a multiple of $n(x)$ for $v,w \in M_x$, we conclude that

$$dA(v,w) = |v \times w|$$

if $[v,w] = \mu_x$. If we wish to compute the area of M, we must evaluate $\int_{[0,1]^2} c^*(dA)$ for orientation-preserving singular 2-cubes c. Define

$$E(a) = [D_1c^1(a)]^2 + [D_1c^2(a)]^2 + [D_1c^3(a)]^2,$$

$$F(a) = D_1c^1(a) \cdot D_2c^1(a) + D_1c^2(a) \cdot D_2c^2(a) + D_1c^3(a) \cdot D_2c^3(a),$$

$$G(a) = [D_2c^1(a)]^2 + [D_2c^2(a)]^2 + [D_2c^3(a)]^2.$$

Then

$$\begin{aligned} c^*(dA)((e_1)_a,(e_2)_a) &= dA(c_*((e_1)_a),c_*((e_2)_a)) \\ &= |(D_1c^1(a),D_1c^2(a),D_1c^3(a)) \times (D_2c^1(a),D_2c^2(a),D_2c^3(a))| \\ &= \sqrt{E(a)G(a) - F(a)^2} \end{aligned}$$

by Problem 4-9. Thus

$$\int_{[0,1]^2} c^*(dA) = \int_{[0,1]^2} \sqrt{EG - F^2}.$$

Calculating surface area is clearly a foolhardy enterprise; fortunately one seldom needs to know the area of a surface. Moreover, there is a simple expression for dA which suffices for theoretical considerations.

5-6 Theorem. *Let M be an oriented two-dimensional manifold (or manifold-with-boundary) in $\mathbf{R}^3$ and let n be the unit outward normal. Then*

$$(1) \qquad dA = n^1\, dy \wedge dz + n^2\, dz \wedge dx + n^3\, dx \wedge dy.$$

Moreover, on M we have

$$(2) \qquad n^1\, dA = dy \wedge dz.$$
$$(3) \qquad n^2\, dA = dz \wedge dx.$$
$$(4) \qquad n^3\, dA = dx \wedge dy.$$

Proof.

Equation (1) is equivalent to the equation

$$dA(v,w) = \det \begin{pmatrix} v \\ w \\ n(x) \end{pmatrix}.$$

This is seen by expanding the determinant by minors along the bottom row. To prove the other equations, let $z \in \mathbf{R}^3{}_x$. Since $v \times w = \alpha n(x)$ for some $\alpha \in \mathbf{R}$, we have

$$\langle z,n(x)\rangle \cdot \langle v \times w,\, n(x)\rangle = \langle z,n(x)\rangle\alpha = \langle z,\alpha n(x)\rangle = \langle z,v \times w\rangle.$$

Choosing $z = e_1$, e_2, and e_3 we obtain (2), (3), and (4). ∎

A word of caution: if $\omega \in \Lambda^2(\mathbf{R}^3{}_a)$ is defined by

$$\omega = n^1(a) \cdot dy(a) \wedge dz(a) + n^2(a) \cdot dz(a) \wedge dx(a) + n^3(a) \cdot dx(a) \wedge dy(a),$$

it is *not* true, for example, that

$$n^1(a) \cdot \omega = dy(a) \wedge dz(a).$$

The two sides give the same result only when applied to $v,w \in M_a$.

A few remarks should be made to justify the definition of length and surface area we have given. If $c\colon [0,1] \to \mathbf{R}^n$ is differentiable and $c([0,1])$ is a one-dimensional manifold-with-boundary, it can be shown, but the proof is messy, that the length of $c([0,1])$ is indeed the least upper bound of the lengths

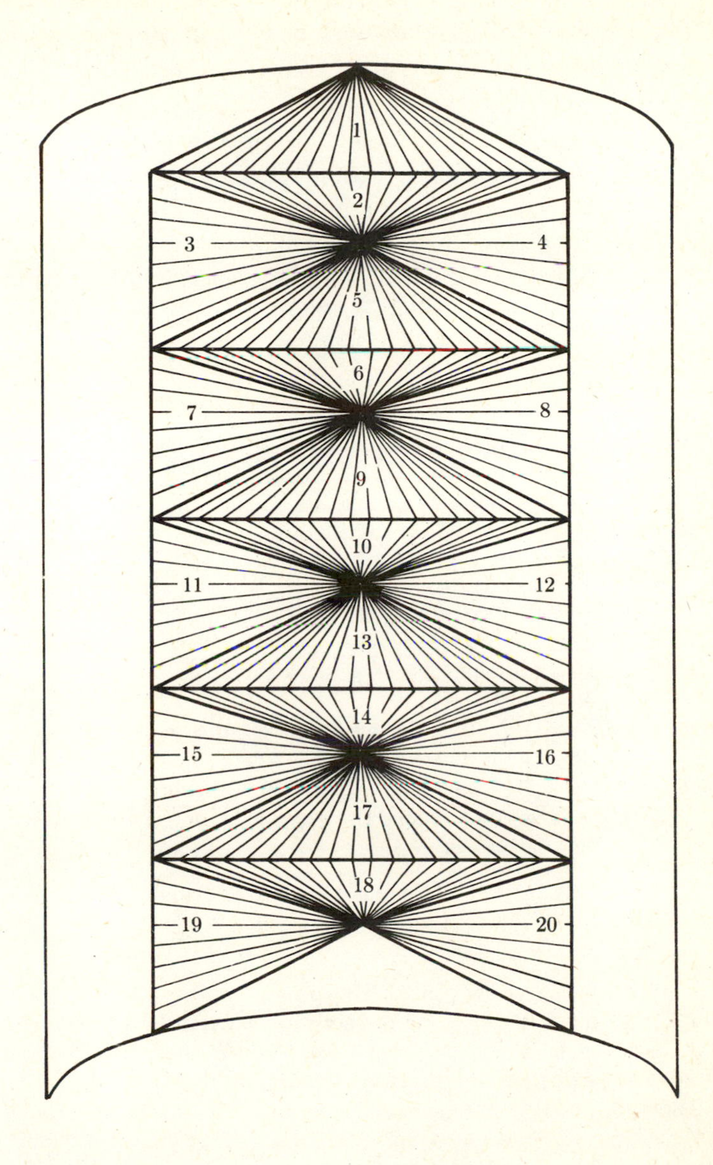

FIGURE 5-9. *A surface containing* 20 *triangles inscribed in a portion of a cylinder. If the number of triangles is increased sufficiently, by making the bases of triangles* 3, 4, 7, 8, *etc., sufficiently small, the total area of the inscribed surface can be made as large as desired.*

of inscribed broken lines. If $c\colon [0,1]^2 \to \mathbf{R}^n$, one naturally hopes that the area of $c([0,1]^2)$ will be the least upper bound of the areas of surfaces made up of triangles whose vertices lie in $c([0,1]^2)$. Amazingly enough, such a least upper bound is usually nonexistent—one can find inscribed polygonal surfaces arbitrarily close to $c([0,1]^2)$ with arbitrarily large area! This is indicated for a cylinder in Figure 5-9. Many definitions of surface area have been proposed, disagreeing with each other, but all agreeing with our definition for differentiable surfaces. For a discussion of these difficult questions the reader is referred to References [3] or [15].

Problems. **5-23.** If M is an oriented one-dimensional manifold in $\mathbf{R}^n$ and $c\colon [0,1] \to M$ is orientation-preserving, show that

$$\int_{[0,1]} c^*(ds) = \int_{[0,1]} \sqrt{[(c^1)']^2 + \cdots + [(c^n)']^2}.$$

5-24. If M is an n-dimensional manifold in $\mathbf{R}^n$, with the usual orientation, show that $dV = dx^1 \wedge \cdots \wedge dx^n$, so that the volume of M, as defined in this section, is the volume as defined in Chapter 3. (Note that this depends on the numerical factor in the definition of $\omega \wedge \eta$.)

5-25. Generalize Theorem 5-6 to the case of an oriented $(n-1)$-dimensional manifold in $\mathbf{R}^n$.

5-26. (a) If $f\colon [a,b] \to \mathbf{R}$ is non-negative and the graph of f in the xy-plane is revolved around the x-axis in $\mathbf{R}^3$ to yield a surface M, show that the area of M is

$$\int_a^b 2\pi f \sqrt{1 + (f')^2}.$$

(b) Compute the area of S^2.

5-27. If $T\colon \mathbf{R}^n \to \mathbf{R}^n$ is a norm preserving linear transformation and M is a k-dimensional manifold in $\mathbf{R}^n$, show that M has the same volume as $T(M)$.

5-28. (a) If M is a k-dimensional manifold, show that an absolute k-tensor $|dV|$ can be defined, even if M is not orientable, so that the volume of M can be defined as $\int_M |dV|$.

(b) If $c\colon [0,2\pi] \times (-1,1) \to \mathbf{R}^3$ is defined by $c(u,v) =$

$$(2\cos u + v\sin(u/2)\cos u,\ 2\sin u + v\sin(u/2)\sin u,\ v\cos u/2),$$

show that $c([0,2\pi] \times (-1,1))$ is a Möbius strip and find its area.

5-29. If there is a nowhere-zero k-form on a k-dimensional manifold M, show that M is orientable.

5-30. (a) If $f\colon [0,1] \to \mathbf{R}$ is differentiable and $c\colon [0,1] \to \mathbf{R}^2$ is defined by $c(x) = (x,f(x))$, show that $c([0,1])$ has length $\int_0^1 \sqrt{1 + (f')^2}$.

(b) Show that this length is the least upper bound of lengths of inscribed broken lines. *Hint:* If $0 = t_0 \le t_1 \le \cdots \le t_n = 1$, then

$$\begin{aligned} |c(t_i) - c(t_{i-1})| &= \sqrt{(t_i - t_{i-1})^2 + (f(t_i) - f(t_{i-1}))^2} \\ &= \sqrt{(t_i - t_{i-1})^2 + f'(s_i)^2(t_i - t_{i-1})^2} \end{aligned}$$

for some $s_i \in [t_{i-1},t_i]$.

5-31. Consider the 2-form ω defined on $\mathbf{R}^3 - 0$ by

$$\omega = \frac{x\,dy \wedge dz + y\,dz \wedge dx + z\,dx \wedge dy}{(x^2 + y^2 + z^2)^{\frac{3}{2}}}.$$

(a) Show that ω is closed.

(b) Show that

$$\omega(p)(v_p,w_p) = \frac{\langle v \times w, p\rangle}{|p|^3}.$$

For $r > 0$ let $S^2(r) = \{x \in \mathbf{R}^3\colon |x| = r\}$. Show that ω restricted to the tangent space of $S^2(r)$ is $1/r^2$ times the volume element, and that $\int_{S^2(r)} \omega = 4\pi$. Conclude that ω is not exact. Nevertheless we denote ω by $d\Theta$ since, as we shall see, $d\Theta$ is the analogue of the 1-form $d\theta$ on $\mathbf{R}^2 - 0$.

(c) If v_p is a tangent vector such that $v = \lambda p$ for some $\lambda \in \mathbf{R}$ show that $d\Theta(p)(v_p,w_p) = 0$ for all w_p. If a two-dimensional manifold M in $\mathbf{R}^3$ is part of a generalized cone, that is, M is the union of segments of rays through the origin, show that $\int_M d\Theta = 0$.

(d) Let $M \subset \mathbf{R}^3 - 0$ be a compact two-dimensional manifold-with-boundary such that every ray through 0 intersects M at most once (Figure 5-10). The union of those rays through 0 which intersect M, is a solid cone $C(M)$. The **solid angle** subtended by M is defined as the area of $C(M) \cap S^2$, or equivalently as $1/r^2$ times the area of $C(M) \cap S^2(r)$ for $r > 0$. Prove that the solid angle subtended by M is $\left|\int_M d\Theta\right|$. *Hint:* Choose r small enough so that there is a three-dimensional manifold-with-boundary N (as in Figure 5-10) such that ∂N is the union of M and $C(M) \cap S^2(r)$, and a part of a generalized cone. (Actually, N will be a manifold-with-corners; see the remarks at the end of the next section.)

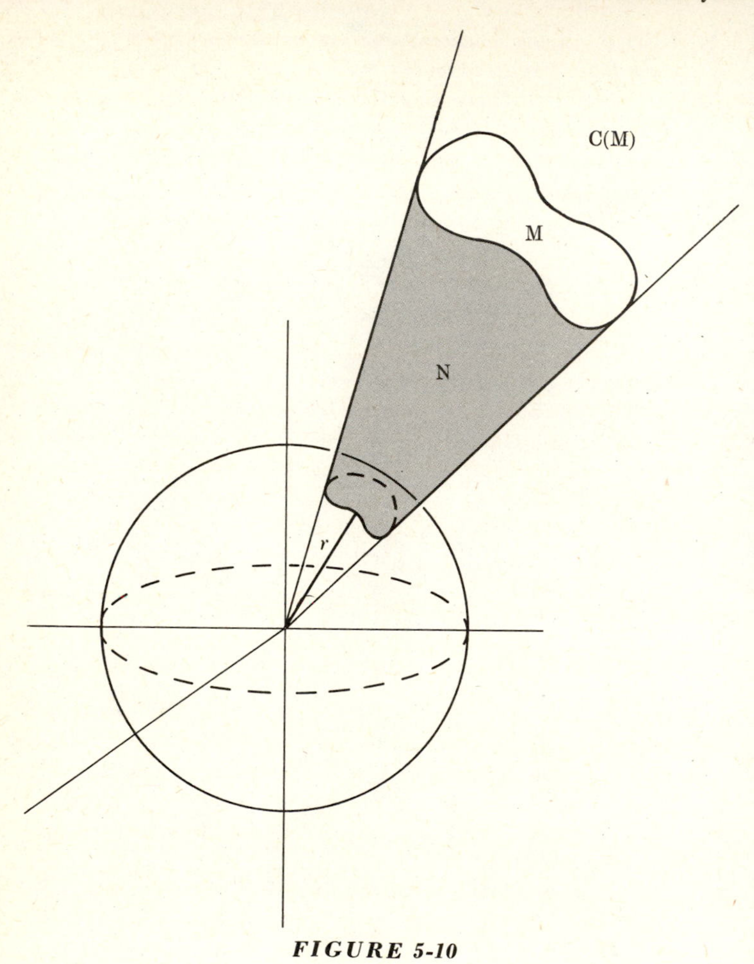

FIGURE 5-10

5-32. Let f, g: $[0,1] \to \mathbf{R}^3$ be nonintersecting closed curves. Define the *linking number* $l(f,g)$ of f and g by (*cf.* Problem 4-34)

$$l(f,g) = \frac{-1}{4\pi} \int_{c_{f,g}} d\Theta.$$

(a) Show that if (F,G) is a homotopy of nonintersecting closed curves, then $l(F_0,G_0) = l(F_1,G_1)$.

(b) If $r(u,v) = |f(u) - g(v)|$ show that

$$l(f,g) = \frac{-1}{4\pi} \int_0^1 \int_0^1 \frac{1}{[r(u,v)]^3} \cdot A(u,v)\, du\, dv$$

where

$$A(u,v) = \det \begin{pmatrix} (f^1)'(u) & (f^2)'(u) & (f^3)'(u) \\ (g^1)'(v) & (g^2)'(v) & (g^3)'(v) \\ f^1(u) - g^1(v) & f^2(u) - g^2(v) & f^3(u) - g^3(v) \end{pmatrix}.$$

(c) Show that $l(f,g) = 0$ if f and g both lie in the xy-plane. The curves of Figure 4-5 (b) are given by $f(u) = (\cos u, \sin u, 0)$ and $g(v) = (1 + \cos v, 0, \sin v)$. You may easily convince yourself that calculating $l(f,g)$ by the above integral is hopeless in this case. The following problem shows how to find $l(f,g)$ without explicit calculations.

5-33. (a) If $(a,b,c) \in \mathbf{R}^3$ define

$$d\Theta_{(a,b,c)} = \frac{(x-a)dy \wedge dz + (y-b)dz \wedge dx + (z-c)dx \wedge dy}{[(x-a)^2 + (y-b)^2 + (z-c)^2]^{\frac{3}{2}}}.$$

If M is a compact two-dimensional manifold-with-boundary in $\mathbf{R}^3$ and $(a,b,c) \notin M$ define

$$\Omega(a,b,c) = \int_M d\Theta_{(a,b,c)}.$$

Let (a,b,c) be a point on the same side of M as the outward normal and (a',b',c') a point on the opposite side. Show that by choosing (a,b,c) sufficiently close to (a',b',c') we can make $\Omega(a,b,c) - \Omega(a',b',c')$ as close to -4π as desired. *Hint:* First show that if $M = \partial N$ then $\Omega(a,b,c) = -4\pi$ for $(a,b,c) \in N - M$ and $\Omega(a,b,c) = 0$ for $(a,b,c) \notin N$.

(b) Suppose $f([0,1]) = \partial M$ for some compact oriented two-dimensional manifold-with-boundary M. (If f does not intersect itself such an M always exists, even if f is knotted, see [6], page 138.) Suppose that whenever g intersects M at x the tangent vector v of g is not in M_x. Let n^+ be the number of intersections where v points in the same direction as the outward normal and n^- the number of other intersections. If $n = n^+ - n^-$ show that

$$n = \frac{-1}{4\pi} \int_g d\Omega.$$

(c) Prove that

$$D_1\Omega(a,b,c) = \int_f \frac{(y-b)dz - (z-c)dy}{r^3}$$

$$D_2\Omega(a,b,c) = \int_f \frac{(z-c)dx - (x-a)dz}{r^3}$$

$$D_3\Omega(a,b,c) = \int_f \frac{(x-a)dy - (y-b)dx}{r^3},$$

where $r(x,y,z) = |(x,y,z)|$.

(d) Show that the integer n of (b) equals the integral of Problem 5-32(b), and use this result to show that $l(f,g) = 1$ if f and g are the curves of Figure 4-6 (b), while $l(f,g) = 0$ if f and g are the curves of Figure 4-6 (c). (These results were known to Gauss [7]. The proofs outlined here are from [4] pp. 409–411; see also [13], Volume 2, pp. 41–43.)

THE CLASSICAL THEOREMS

We have now prepared all the machinery necessary to state and prove the classical "Stokes' type" of theorems. We will indulge in a little bit of self-explanatory classical notation.

5-7 Theorem (Green's Theorem). *Let $M \subset \mathbf{R}^2$ be a compact two-dimensional manifold-with-boundary. Suppose that $\alpha,\beta\colon M \to \mathbf{R}$ are differentiable. Then*

$$\int_{\partial M} \alpha\,dx + \beta\,dy = \int_M (D_1\beta - D_2\alpha)dx \wedge dy$$

$$= \iint_M \left(\frac{\partial\beta}{\partial x} - \frac{\partial\alpha}{\partial y}\right) dx\,dy.$$

(Here M is given the usual orientation, and ∂M the induced orientation, also known as the counterclockwise orientation.)

Proof. This is a very special case of Theorem 5-5, since $d(\alpha\,dx + \beta\,dy) = (D_1\beta - D_2\alpha)dx \wedge dy$. ▌

5-8 Theorem (Divergence Theorem). *Let $M \subset \mathbf{R}^3$ be a compact three-dimensional manifold-with-boundary and n the unit outward normal on ∂M. Let F be a differentiable vector field on M. Then*

$$\int_M \operatorname{div} F\, dV = \int_{\partial M} \langle F, n \rangle\, dA.$$

This equation is also written in terms of three differentiable functions $\alpha, \beta, \gamma \colon M \to \mathbf{R}$:

$$\iiint_M \left(\frac{\partial \alpha}{\partial x} + \frac{\partial \beta}{\partial y} + \frac{\partial \gamma}{\partial z} \right) dV = \iint_{\partial M} (n^1 \alpha + n^2 \beta + n^3 \gamma)\, dS.$$

Proof. Define ω on M by $\omega = F^1\, dy \wedge dz + F^2\, dz \wedge dx + F^3\, dx \wedge dy$. Then $d\omega = \operatorname{div} F\, dV$. According to Theorem 5-6, on ∂M we have

$$\begin{aligned} n^1\, dA &= dy \wedge dz, \\ n^2\, dA &= dz \wedge dx, \\ n^3\, dA &= dx \wedge dy. \end{aligned}$$

Therefore on ∂M we have

$$\begin{aligned} \langle F, n \rangle\, dA &= F^1 n^1\, dA + F^2 n^2\, dA + F^3 n^3\, dA \\ &= F^1\, dy \wedge dz + F^2\, dz \wedge dx + F^3\, dx \wedge dy \\ &= \omega. \end{aligned}$$

Thus, by Theorem 5-5 we have

$$\int_M \operatorname{div} F\, dV = \int_M d\omega = \int_{\partial M} \omega = \int_{\partial M} \langle F, n \rangle\, dA. \quad \blacksquare$$

5-9 Theorem (Stokes' Theorem). *Let $M \subset \mathbf{R}^3$ be a compact oriented two-dimensional manifold-with-boundary and n the unit outward normal on M determined by the orientation of M. Let ∂M have the induced orientation. Let T be the vector field on ∂M with $ds(T) = 1$ and let F be a differentiable vector field in an open set containing M. Then*

$$\int_M \langle (\nabla \times F), n \rangle\, dA = \int_{\partial M} \langle F, T \rangle\, ds.$$

This equation is sometimes written

$$\int_{\partial M} \alpha\,dx + \beta\,dy + \gamma\,dz =$$

$$\iint_M \left[n^1\left(\frac{\partial\gamma}{\partial y} - \frac{\partial\beta}{\partial z}\right) + n^2\left(\frac{\partial\alpha}{\partial z} - \frac{\partial\gamma}{\partial x}\right) + n^3\left(\frac{\partial\beta}{\partial x} - \frac{\partial\alpha}{\partial y}\right)\right] dS.$$

Proof. Define ω on M by $\omega = F^1\,dx + F^2\,dy + F^3\,dz$. Since $\nabla \times F$ has components $D_2F^3 - D_3F^2$, $D_3F^1 - D_1F^3$, $D_1F^2 - D_2F^1$, it follows, as in the proof of Theorem 5-8, that on M we have

$$\begin{aligned}\langle(\nabla \times F), n\rangle\,dA &= (D_2F^3 - D_3F^2)dy \wedge dz \\ &\qquad + (D_3F^1 - D_1F^3)dz \wedge dx \\ &\qquad\qquad + (D_1F^2 - D_2F^1)dx \wedge dy \\ &= d\omega.\end{aligned}$$

On the other hand, since $ds(T) = 1$, on ∂M we have

$$\begin{aligned}T^1\,ds &= dx,\\ T^2\,ds &= dy,\\ T^3\,ds &= dz.\end{aligned}$$

(These equations may be checked by applying both sides to T_x, for $x \in \partial M$, since T_x is a basis for $(\partial M)_x$.)

Therefore on ∂M we have

$$\begin{aligned}\langle F,T\rangle\,ds &= F^1T^1\,ds + F^2T^2\,ds + F^3T^3\,ds\\ &= F^1\,dx + F^2\,dy + F^3\,dz\\ &= \omega.\end{aligned}$$

Thus, by Theorem 5-5, we have

$$\int_M \langle(\nabla \times F),n\rangle\,dA = \int_M d\omega = \int_{\partial M} \omega = \int_{\partial M} \langle F,T\rangle\,ds. \quad ∎$$

Theorems 5-8 and 5-9 are the basis for the names div F and curl F. If $F(x)$ is the velocity vector of a fluid at x (at some time) then $\int_{\partial M} \langle F,n\rangle\,dA$ is the amount of fluid "diverging" from M. Consequently the condition div $F = 0$ expresses the fact that the fluid is incompressible. If M is a disc, then $\int_{\partial M} \langle F,T \rangle \, ds$ measures the amount that the fluid curls around the center of the disc. If this is zero for all discs, then $\nabla \times F = 0$, and the fluid is called *irrotational.*

These interpretations of div F and curl F are due to Maxwell [13]. Maxwell actually worked with the negative of div F, which he accordingly called the *convergence.* For $\nabla \times F$ Maxwell proposed "with great diffidence" the terminology *rotation of F;* this unfortunate term suggested the abbreviation rot F which one occasionally still sees.

The classical theorems of this section are usually stated in somewhat greater generality than they are here. For example, Green's Theorem is true for a square, and the Divergence Theorem is true for a cube. These two particular facts can be proved by approximating the square or cube by manifolds-with-boundary. A thorough generalization of the theorems of this section requires the concept of manifolds-with-corners; these are subsets of $\mathbf{R}^n$ which are, up to diffeomorphism, locally a portion of $\mathbf{R}^k$ which is bounded by pieces of $(k-1)$-planes. The ambitious reader will find it a challenging exercise to define manifolds-with-corners rigorously and to investigate how the results of this entire chapter may be generalized.

Problems. **5-34.** Generalize the divergence theorem to the case of an n-manifold with boundary in $\mathbf{R}^n$.

5-35. Applying the generalized divergence theorem to the set $M = \{x \in \mathbf{R}^n\colon |x| \le a\}$ and $F(x) = x_x$, find the volume of $S^{n-1} = \{x \in \mathbf{R}^n\colon |x| = 1\}$ in terms of the n-dimensional volume of $B_n = \{x \in \mathbf{R}^n\colon |x| \le 1\}$. (This volume is $\pi^{n/2}/(n/2)!$ if n is even and $2^{(n+1)/2}\pi^{(n-1)/2}/1 \cdot 3 \cdot 5 \cdot \ldots \cdot n$ if n is odd.)

5-36. Define F on $\mathbf{R}^3$ by $F(x) = (0,0,cx^3)_x$ and let M be a compact three-dimensional manifold-with-boundary with $M \subset \{x\colon x^3 \le 0\}$. The vector field F may be thought of as the downward pressure of a fluid of density c in $\{x\colon x^3 \le 0\}$. Since a fluid exerts equal pressures in all directions, we define the *buoyant force* on M, due to the fluid, as $-\int_{\partial M} \langle F,n \rangle \, dA$. Prove the following theorem.
Theorem (Archimedes). The buoyant force on M is equal to the weight of the fluid displaced by M.

Bibliography

1. Ahlfors, *Complex Analysis*, McGraw-Hill, New York, 1953.
2. Auslander and MacKenzie, *Introduction to Differentiable Manifolds*, McGraw-Hill, New York, 1963.
3. Cesari, *Surface Area*, Princeton University Press, Princeton, New Jersey, 1956.
4. Courant, *Differential and Integral Calculus*, Volume II, Interscience, New York, 1937.
5. Dieudonné, *Foundations of Modern Analysis*, Academic Press, New York, 1960.
6. Fort, *Topology of 3-Manifolds*, Prentice-Hall, Englewood Cliffs, New Jersey, 1962.
7. Gauss, *Zur mathematischen Theorie der electrodynamischen Wirkungen*, [4] (Nachlass) Werke V, 605.
8. Helgason, *Differential Geometry and Symmetric Spaces*, Academic Press, New York, 1962.
9. Hilton and Wylie, *Homology Theory*, Cambridge University Press, New York, 1960.
10. Hu, *Homotopy Theory*, Academic Press, New York, 1959.
11. Kelley, *General Topology*, Van Nostrand, Princeton, New Jersey, 1955.

12. Kobayashi and Nomizu, *Foundations of Differential Geometry*, Interscience, New York, 1963.

13. Maxwell, *Electricity and Magnetism*, Dover, New York, 1954.

14. Natanson, *Theory of Functions of a Real Variable*, Frederick Ungar, New York, 1955.

15. Radó, *Length and Area*, Volume XXX, American Mathematical Society, Colloquium Publications, New York, 1948.

16. de Rham, *Variétés Differentiables*, Hermann, Paris, 1955.

17. Sternberg, *Lectures on Differential Geometry*, Prentice-Hall, Englewood Cliffs, New Jersey, 1964.

Index

Addenda

1. It should be remarked after Theorem 2-11 (the Inverse Function Theorem) that the formula for f^{-1} allows us to conclude that f^{-1} is actually continuously differentiable (and that it is C^∞ if f is). Indeed, it suffices to note that the entries of the inverse of a matrix A are C^∞ functions of the entries of A. This follows from "Cramer's Rule": $(A^{-1})_{ji} = (\det A^{ij})/(\det A)$, where A^{ij} is the matrix obtained from A by deleting row i and column j.

2. The proof of the first part of Theorem 3-8 can be simplified considerably, rendering Lemma 3-7 unnecessary. It suffices to cover B by the interiors of closed rectangles U_i with $\sum_{i=1}^{\infty} v(U_i) < \varepsilon$, and to choose for each $x \in A - B$ a closed rectangle V_x, containing x in its interior, with $M_{V_x}(f) - m_{V_x}(f) < \varepsilon$. If every subrectangle of a partition P is contained in one of some finite collection of U_i's and V_x's which cover A, and $|f(x)| \leq M$ for all x in A, then $U(f, P) - L(f, P) < \varepsilon v(A) + 2M\varepsilon$.

The proof of the converse part contains an error, since $M_S(f) - m_S(f) \geq 1/n$ is guaranteed only if the interior of S intersects $B_{1/n}$. To compensate for this it suffices to cover the boundaries of all subrectangles of P with a finite collection of rectangles with total volume $< \varepsilon$. These, together with $\mathcal{S}$, cover $B_{1/n}$, and have total volume $< 2\varepsilon$.

3. The argument in the first part of Theorem 3-14 (Sard's Theorem) requires a little amplification. If $U \subset A$ is a closed rectangle with sides of length l, then, because U is compact, there is an integer N with the following property: if U is divided into N^n rectangles, with sides of length l/N, then $|D_jg^i(w) - D_jg^i(z)| < \varepsilon/n^2$ whenever w and z are both in one such rectangle S. Given $x \in S$, let $f(z) = Dg(x)(z) - g(z)$. Then, if $z \in S$,

$$|D_jf^i(z)| = |D_jg^i(x) - D_jg^i(z)| < \varepsilon/n^2.$$

So by Lemma 2-10, if $x,y \in S$, then

$$\begin{aligned} |Dg(x)(y - x) - g(y) + g(x)| = |f(y) - f(x)| &< \varepsilon|x - y| \\ &\leq \varepsilon\sqrt{n}\,(l/N). \end{aligned}$$

4. Finally, the notation $\Lambda^k(V)$ appearing in this book is incorrect, since it conflicts with the standard definition of $\Lambda^k(V)$ (as a certain quotient of the tensor algebra of V). For the vector space in question (which is naturally isomorphic to $\Lambda^k(V^*)$ for finite dimensional vector spaces V) the notation $\Omega^k(V)$ is probably on the way to becoming standard. This substitution should be made on pages 78–85, 88–89, 116, and 126–128.